THE BEAST OF BOGGY CREEK

THE TRUE STORY OF THE FOUKE MONSTER

Lyle Blackburn

AПOMALIST BOOKS
San Antonio * Charlottesville

An Original Publication of ANOMALIST BOOKS
The Beast of Boggy Creek: The True Story of the Fouke Monster
Copyright © 2012 by Lyle Blackburn
ISBN: 1933665572

Cover artwork and layout by Justin Osbourn of Slasher Design
(www.osbourndraw.com)

Sighting illustrations by Dan Brereton (www.nocturnals.com)

Map illustrations by Lyle Blackburn

Miller County Historical Society photos courtesy of Frank McFerrin

Other photos courtesy of individual photographers as credited

Book design by Seale Studios

Grateful acknowledgement is made for permission to reprint material
from…

- Various newsprint articles. By permission of the *Texarkana Gazette*.
- Various sighting reports and articles from the Texas Bigfoot Research Conservancy website. By permission of Daryl Colyer and Alton Higgins.
- Various sighting reports from the Gulf Coast Bigfoot Research Organization website. By permission of Bobby Hamilton.

For more information about the author, visit: www.monstrobizarro.com

For the latest on the Fouke Monster, visit: www.foukemonster.net

For information about Anomalist Books, visit anomalistbooks.com, or
write to: 5150 Broadway #108, San Antonio, TX 78209

For my grandmother, Bette Capps, who always believes in every crazy thing I do.

CONTENTS

The most beautiful thing we can experience is the mysterious.
It is the source of all true art and all science.
—Albert Einstein

FOREWORD:
Why Are Fouke's Boggy Creek Creatures So Important?

Something shocking and historically important took place in 1972. A drive-in Bigfoot movie became a surprise moneymaker. The movie was *The Legend of Boggy Creek*, released in 1972, and out for the first time on DVD in 2002.

Despite the fact that people around the Boggy Creek area had been seeing Swamp Ape type creatures since at least the 1940s, their encounters in the 1960s and then especially in 1971, received a good deal of media attention. As Lyle Blackburn's account makes clear in great detail, the large and hairy "Fouke Monsters" gained notoriety when one or more unknown hominoids harassed two families (the Fords and the Crabtrees) living outside Fouke, Arkansas (population 600), in the southwest part of the state. Director Charles B. Pierce decided to use real eyewitnesses and the actual locations near Boggy Creek to recreate Fouke's experience with their local monster. The docudrama or semi-documentary thriller became a smash success, a cult classic.

Although a scripted movie, the spooky footage of the river bottoms, fog, and vegetation along Boggy Creek made for a captivating, and for most filmgoers, scary setting. I am constantly struck by fellow researchers and members of the general public who tell me that it was *this* movie that got them involved in the pursuit of more information on Bigfoot and other cryptids.

The impact of *The Legend of Boggy Creek* has been far-reaching. A couple of modern reviews from the internet give more than a hint of its significance: "Bigfoot was, and still is, a celebrity because of this movie!" and "This may be the movie that made 'Bigfoot' a national star."

A self-published book by Smokey Crabtree entitled *Smokey and the Fouke Monster* (1974) followed the film, giving "another point of view" of the events portrayed in the Boggy Creek movie. Smokey and I lectured from the same stage in Ohio years ago, and he's still talking as if it all happened yesterday. Crabtree never saw the Fouke Monster, but the movie still changed his life.

The movie also created a whole new generation of dedicated Bigfoot hunters. Young people between the ages of 10 and 13 who were first attracted to Bigfoot research in the 1970s, speak of *The Legend of Boggy Creek* as the source of their passion in the subject. In his 1988 book, *Big Footnotes*, Daniel Perez wrote: "My personal interest in monsters was first ignited at about the tender age of 10, by the movie *The Legend of Boggy Creek*. This was the trigger which lead to casual to casually serious to serious full-fledged involvement in this subject matter." Maryland's *Bigfoot Digest* author Mark Opsasnick notes this movie inspired his interest in Bigfoot at the age of 11. Ditto for cryptozoology artist Bill Rebsamen, who told me, "I was about 10 years old when I saw it. I went immediately to the library the next day and checked out all the books I could find on Bigfoot." And Chester Moore, Jr., Texan outdoors journalist and author of *Bigfoot South* (2002), writes: "Seeing *The Legend of Boggy Creek* lit my interest in the Bigfoot phenomenon into a full-blown passion. While the Pacific Northwest seemed a world away to me, Arkansas did not…The impact it had on me as a youngster was immense."

Please note, however, these Hollywood Swamp Ape and Bigfoot movies did not lead to a rash of Boggy Creek-type creature sightings in most places in North America. No, instead, what it did do was stimulate and influence future researchers to be open-minded about the possibilities of such unknown beasts being out there.

—**Loren Coleman**, director of the International Cryptozoology Museum and author of *Mysterious America, Bigfoot! The True Story of Apes in America, The Field Guide to Bigfoot,* and other cryptozoology books.

INTRODUCTION

For more than a century, tales of a mysterious ape-like creature lurking in the woods of southern Arkansas have circulated among believers and skeptics alike. The Fouke Monster, as it is called, has become one of those enduring artifacts of backwoods legend, fueled by news reports, movies, internet, and cryptozoological studies until it has earned a solid foothold within American lore. To those who believe to have seen it, it is real; to the skeptical, it is simply a campfire story; to Hollywood, a bankroll; and to those with a love for monsters or local lore, it is a subject worthy of continued research. But regardless of your affiliation, there is something interesting for all in the tale of the Fouke Monster. That's because it is more than just the simple story of a monster. It is an exploration into primal fears, cultural phenomenon, cryptozoology, and the magic of movies all rolled into one.

Like many, my first exposure to the Fouke Monster (pronounced *Fowk*) came in the form of celluloid cinema with the movie *The Legend of Boggy Creek*, originally released in 1972. This pseudo-documentary directed by Charles B. Pierce gave national attention to the small town of Fouke, Arkansas, whose namesake monster would be propelled into the pantheon of undocumented creatures alongside Bigfoot, Loch Ness Monster, Mothman, and others. I was fortunate to catch *The Legend of Boggy Creek* in re-run at an old drive-in as a child. Having grown up near the movie's real-life setting in Fouke (about three hours drive from my home in Fort Worth, Texas), it hit *very* close to home when I first heard the creature's scream during the opening sequence. I was familiar with the backwoods of the Texas/Arkansas area—my father was a bowhunter who didn't mind dragging his young son along on twilight stakeouts to hunt the local game—so it was not a stretch for me to imagine a seven-foot ape-beast lurking just out of sight on those crisp autumn nights. The movie not only scared me, but furthered my love for unexplained creatures and crowned the Fouke Monster as my very own homegrown beastie.[1]

As I grew up, my fascination with the Beast of Boggy Creek graduated from childlike wonder to legitimate scientific speculation, eventually encompassing not only a passion for the "monster" but also an interest in the social frenzy that has surrounded it. To me, the tale represents a microcosm of what makes up the most interesting aspects of world mysteries, social interaction, and old time folktales: the phenomenon of instinctual fear running headfirst into mass media madness.

After being reunited with *The Legend of Boggy Creek* when it was re-released several years ago on DVD, and with the benefit of the internet, I spent time poking around trying to find out more information about the "truth" as referenced in the movie's tagline: "A True Story." To my disappointment there was not a comprehensive website, book, or even a magazine article that would tie together the complete and fascinating tale. There were only pieces that could be gathered by a smattering of old news articles, cryptozoological summaries, and regurgitated movie reviews. It seemed the only significant information was to be found in three self-published autobiographies by Fouke local, J.E. "Smokey" Crabtree, who writes of the monster, but only as it pertains to his own life story.

Coming up short in my quest for the complete story, I decided to take matters into my own hands and the idea for this book was born. It was time for the Fouke Monster to receive its own official chronicle. After all, the little-known Lake Worth Monster (of Fort Worth, Texas fame) has its own book, Nessie and Champ have theirs, and even the mongrel Chupacabra has recently come into its own, but time was mostly leaving the Fouke Monster behind.

Even more disturbing, over the years the Fouke Monster's legend has been passed around in an often incomplete fashion, leaving out relevant information or mixing the true and sensational aspects together as one—not unlike what the seminal movie had done. So my first goal was to gather all the information I could find and attempt to sort fact from fiction. As I did so, I began to realize that the full impact of the monster's story could not be felt without also exploring the rich history of the area, the cast of characters, and the hidden secrets of the town that shares the burden of notoriety.

After all, the tendrils of lasting legends inevitably root deep within the culture from which they are born.

Unlike the ubiquitous tales of Bigfoot, which are not connected to a single town (and in fact span across state lines and even continents), the "monster" of *this* story is forever tied to a particular location in both name and tradition. It is defined not only by its distinct Southern heritage, but by the locals who tell its story. It is therefore not surprising that its description at times reflects the hot, sticky, shadowy Arkansas swamplands from which it has come. Unkempt, rebellious, and malodorous, it seems to have an extra dimension of personality not found in the average, faceless man-like apes that are said to roam the majestic forests elsewhere. For me, it is this connection to the local small town culture and the inhospitable swamps that makes the story of the Fouke Monster interesting well beyond the simple affection for a mysterious creature, and another reason why I have enjoyed writing this book.

I must point out that this book is not an attempt to ultimately prove or disprove the existence of the Fouke Monster (and its kind, since logically there would be more than one). I will leave the considerable burden of proof in the hands of more qualified individuals such as scientists or wildlife biologists. While I have paid close attention to providing accurate and factual details, my main purpose here is to provide an entertaining and comprehensive account of the creature, discuss some of the popular theories, and perhaps along the way throw out an opinion or two based on my observations. But these are just personal opinions, as no doubt you will have some of your own.

So as you follow me through the intriguing tale of this humble cryptid as it rises from the swamp to the big screen and back again, I hope you will enjoy reading as much as I have enjoyed writing about it. Real or not, I will never forget when I first heard those opening words spoken by the narrator of *The Legend of Boggy Creek* as the movie flashed to life on that enormous drive-in screen so many years ago: "I was 7 years old when I first heard him scream. It scared me then and it scares me now."

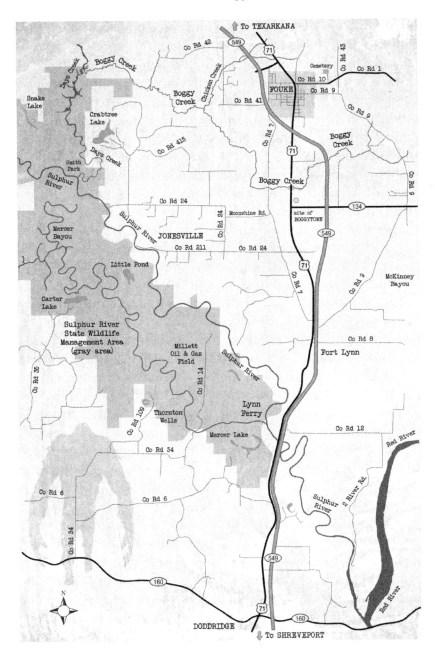

Fouke, Arkansas and the surrounding area.

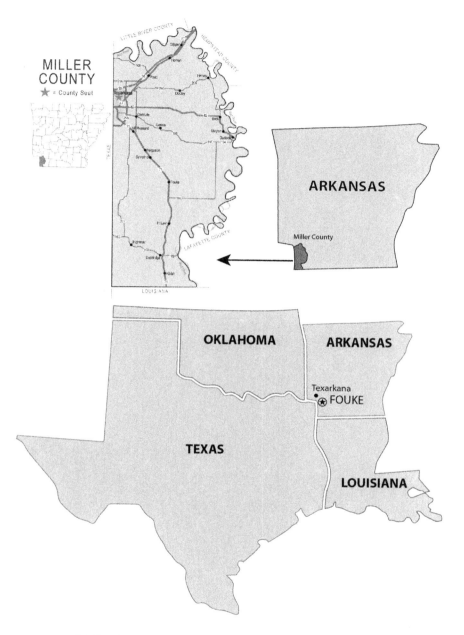

Miller County and the neighboring four-state region.

1. SETTING THE STAGE

THE QUIET BEFORE THE STORM

The year was 1971. A storm of change and progress was brewing in the skies over America. President Richard M. Nixon and his administration struggled with the inherited burden of the Vietnam War as the age of the flower child was coming to a close. The free love spell had already been broken by the Charles Manson murders and the Kent State massacre the previous year, and soon it would draw its final breath with the mysterious passing of poet rock icon Jim Morrison of The Doors, on July 3, 1971. The Doors newly released single, "Riders on the Storm," was eerily symbolic of the times with its ominous lyrics, spooky backbeat, and thunderstorm overdubs. Inflation was high, tensions were high, and an air of uncertainty permeated the country as its citizens wondered which direction the tempest would take.

Along with the communist turmoil in Vietnam and the underlying Cold War, the space race was still in full stride with Apollo 14 becoming the third successful lunar landing. At home, families forgot about the world's problems as they enjoyed the innocence of television shows such as *The Partridge Family* and *The Odd Couple*, or the crime solving work of *McCloud*. Outside the home, Americans flocked to the newly opened Walt Disney World in Florida and crowded into stadiums to cheer the Dallas Cowboys on their way to Super Bowl VI. The motorcycle daredevil, Evil Knievel, thrilled audiences as he set a new world record by jumping 19 cars. The engines of a new decade were revving.

But in a small Arkansas town called Fouke, things were much quieter. While the world raced around in a storm of jet fuel and politics, the people of this town pretty much did what they had

1

always done since it had been founded nearly a century earlier. They worked hard during the week, went to church on Sundays, placed a high importance on family and friends, and enjoyed good Southern cooking whenever they could. Rural life in the early 1970s was not an easy life by any means, but it was a simpler life.

This is not to say that the Vietnam War passed without effect on the town. It did take away several of the local boys and turned them into soldiers. Some even became hometown heroes. And the town's youth were not completely immune from the temptations of the hippy movement or the trappings of rock music either. It was, after all, a part of normal teenage rebellion. But despite these outside influences, gospel and country were still the order of the day in 1971, keeping the town of 500 residents and the surrounding areas grounded in good ol' small town Americana. Here, everyone knew their neighbors, shared stories, and pitched in to help out when something needed building or fixing. Whenever time permitted, the men headed off into the surrounding woods to hunt or fish, while the women got together for social clubs or church functions. Even though Fouke was only 15 miles south of the large metropolitan area of Texarkana, it maintained a peaceful seclusion separate from the bustle of the city. To the residents of the time, it was nothing short of country paradise.

That is, until May 3, 1971, changed everything.

That was the day the *Texarkana Gazette* printed the first in a series of hair-raising reports about a monster that allegedly haunted the woods near Fouke. The monster was said to be a large, hairy ape-like creature that walked upright on two legs. It stood nearly seven feet tall, had glowing red eyes, gave off a rank odor, and occasionally let out a horrifying shriek. The description was not unlike that of Sasquatch or Bigfoot, but this creature had a decidedly Southern slant in that it seemed to be leaner, meaner, and hairier. As more reports came in, it was apparent that the *thing*—whatever it was— preferred the proximity of Boggy Creek, a ruddy tributary which snakes up and around Fouke like a long, forked tongue.

The incident in early May, as reported by *Texarkana Gazette* writer Jim Powell, involved a late night attack on the Ford residence located on the north side of Fouke. According to five adult witnesses, a large hairy "monster" stuck his hand through an open window and tried to gain entrance into the home through the back door.

Twenty days later, three people driving along Highway 71 at night witnessed "a large hairy creature" run across the road near the Boggy Creek bridge. In the following weeks, a tall, hair-covered creature with long arms was spotted by three more adults in the area.

When these reports hit the newspaper, it not only sent a chill through the residents, but also brought to light other incidents involving the "monster" that had not been widely known up until that point. Apparently, the creature had also been prowling around the homes of residents living near Boggy Creek. One man had even taken a few shots at the creature with his old Army rifle but had never succeeded in bringing it down.

Fouke's constable, Ernest Walraven, pointed out that similar horrifying reports had also circulated years earlier when a creature fitting the same description had been seen in Jonesville, a small community a few miles southwest of Fouke. In one instance, a teenager had come face-to-face with a hairy wildman while squirrel hunting. He fired three times and ran. By the time he made it back home, the boy was frantic. Though many of the townsfolk had heard the story over the years, it had never been made public by way of a news report. It was not something they wished to speak about if they could avoid it. But when the new rash of incidents made headlines in '71, the old stories began to resurface. This only added more fuel to the growing fire. As the weeks went on, it was becoming apparent that a strange animal might be living in the woods near Fouke.

For some locals it was laughable, for others it was frightening. A few were not surprised by the stories as they had heard about the creature for years. But to the outside world it was nothing short of intriguing. In no time, the town was the subject of rumors and gossip, as well as waves of monster hunters, filmmakers, documentarians, and a steady stream of tourists which continues even today.

And throughout all the highs and lows that come with the limelight, sightings of the monster *did not stop*. Some uninformed outsiders may tell you the reports stopped back in the late 1970s, but this is *definitely* not the case. The locals know better. By all accounts, there is still something mysterious living in the woods out there, and it is no less intriguing now than when the creature first came to fame all those years ago. The story is still unfolding. Like a slow Southern drawl, things take their own sweet time in the South.

THE NATURAL STATE

It's hard to imagine a setting more suitable for a monster tale than southern Arkansas, a land populated by thick trees whose gnarled limbs cast misshapen shadows on the swampy bottoms. Here, the buzz of modern American life is easily replaced by the drone of insects, and an occasional howl in the darkness is hard to distinguish as animal, human, or perhaps something more mysterious. This is a long-standing patch of American landscape that has, in many ways, retained the essence of its foundation built by the generations of hardworking people who forged a living from its rich soil. And not unlike other regions of deep-rooted Americana, this land possesses its share of legends and lore that have served to fire the imaginations— and fears—of not only the people of this area, but of others beyond its boundaries.

Arkansas' state slogan, "The Natural State," effectively describes what can be found throughout much of the state today. Acres of unblemished forestry, miles of swampland, caves, lakes, rivers, and other points of natural beauty make up a large portion of the attraction, and bolsters their tourist trade, which lists "outdoors" at the top of the list on state's official tourism website. The state boasts three national forests, which collectively engulf over 2.9 million acres. Toss in 9,000 miles of stream and river waterways, and upwards of 600,000 acres of sprawling lakes, and it leaves little doubt that this land is brimming with wildlife of all sorts, most of which has occupied space in the state far longer than most people realize.

The first record of humans inhabiting the Arkansas area begins about 13,500 years ago during the Pleistocene epoch.[2] Referred to as Paleoindians, these early humans survived as hunter-gatherers forging off the land and hunting game such as deer and rabbits. During this time North America's climate was much colder than it is today. As such, the Paleoindians would have also existed alongside many now-extinct animals, such as the mastodon, giant sloth, and cave bear.

As the climate began to moderate and increase in average temperature, many of the larger mammals became extinct, leaving only the familiar fauna we find today (plus or minus any unknown apes, of course). The forests expanded and the waterways changed course as they twisted across the Arkansas landscape looking for the best seaward routes. One of these waterways was the Red River, which carved out a rich and plentiful basin across the southern corner of Arkansas. This mighty river offered a dependable lifeline for both human and animal alike, making the area desirable for all varieties of inhabitants.

In close proximity to the Red River, and only a few miles southwest of Fouke, lies the vast Sulphur River Bottoms where it has long been conjectured the monster dwells. Deriving its name from the Sulphur River—a long tributary that winds its way from Wright-Patman Lake in Texas, across the corner of Arkansas, and on down into Louisiana—the bottoms include the 18,000 acre Sulphur River Wildlife Management Area, one of the last remaining tracts of untouched habitat along the Red River Valley. The land here is covered with dense patches of hardwoods, cypress breaks, and a spidery network of creeks and oxbow lakes. A tangle of thorns and a variety of poisonous snakes and other dangerous animals add to the unfriendly wetland terrain to make the entire bottomlands extremely inhospitable and remote.

In fact, if there were such a place to hide a throw-back to the time when ape and man occupied a closer proximity on the branches of life, this particular spot must surely rank near the top. Even today, this land contains large pockets of swampland and forestry that are difficult to access. The sparse county roads often dead-end into brackish swamp water where one would need to trade wheels for a sturdy canoe or pirogue in order to continue the journey. Standing waters such as Mercer Bayou extend in all directions from the

The Sulphur River as it winds through an area southwest of Fouke.
(Photo by the author)

Sulphur River, holding the land captive and wholly undesirable for modern renovation. Duckweed grows everywhere like a green carpet over the stagnant bayous, and cypress trees grow thick and hang low, making it difficult to see any local inhabitants who might be hiding beneath the waters or in the long shadows of the crowded flora.

Looking out over the Sulphur River Wildlife Management Area from a high vantage point or from the air, it is easier to grasp just how dense and uninhabited the area is. The tree tops extend for miles in an endless canopy, while the rivers and creekbeds weave through the land like giant watery snakes. If something wanted to hide there, it would not be hard, especially if the species population was thin and it had an affinity for surviving beyond the reach of civilization.

As the Sulphur River winds southward, it eventually converges with the Red River and flows south to join the greater network of

Aerial photo of the Mercer Bayou area.
(Photo by Ken Stewart)

waterways known collectively as the Mississippi Waterweb. Upon its journey, the Red River meanders near other equally spooky areas sprawled along the Louisiana-Texas border, such as Black Lake Bayou, Cypress Bayou, and Caddo Lake. These areas themselves boast a long history of similar "monster" sightings. When considering the relatively close proximity to the town of Fouke, it would seem as though the entire area where the tri-states converge adds up to the kind of place that a monster could only dream of.

But is it more than just dreaming? Do the spooky swamps and riverways *inspire* such monster sightings, or does an unknown creature live here *because* of the remote privacy that the area affords? That remains to be proven, but either way, it is certain that these parts of Arkansas have remained close to their original state when the North American ice age shaped the terrain long ago.

THE ARKANSAS WILD MAN

Throughout history, in all corners of the world, people have reported seeing "hairy wild men" who occasionally emerge from the forests and cross paths with those of the civilized world. Often interwoven into the lore of werewolves or modern day Bigfoot, these wild men have a long history of sightings, which are eerily similar in all cultures. Arkansas is no different, having a few bona fide wild man stories of its own dating back to the time when early Americans began settling the area.

Whether or not these can be attributed to the Fouke Monster — or Sasquatch in general — is, of course, impossible to determine due to the lack of solid evidence and typically sketchy details. However, these accounts are certainly important in that they provide a historical record of hairy creatures being reported in the state long before the notion of Bigfoot became popular in American culture.

As Spaniards began to explore the Arkansas region starting around 1541, they wrote of encounters with Native American tribes such as the Tunica, Caddo, Quapaw, and Osage. In looking at these early tribes and others, we find that they too may have encountered wild men or Sasquatch-like creatures in the area. In fact, most Native American cultures have stories or mythos about some kind of large, hairy man-like creature said to inhabit the forests of North America and British Columbia. Though the stories and representations differ slightly from tribe to tribe, each has a specific word to represent this creature in their language. According to Kathy Moskowitz Strain, archaeologist and author of *Giants, Cannibals & Monsters: Bigfoot in Native Culture*, the Caddos used the word *Ha'yacatsi* to describe such a creature. The word literally translates to "lost giant," which seems consistent with the modern mythos of the Sasquatch. Others include the Cherokee word *Kecleh-Kudleh* (hairy savage), the Creek word *Honka* (hairy man), and the Choctaw words *Kashehotapalo* (cannibal man) and *Nalusa Falaya* (big giant).

Some researchers argue that these entities were nothing more than spirit animals, but others point to evidence suggesting that they were real living, breathing creatures. For example, masks and totem poles made by these early Americans depict faces that are eerily similar to those of apes. If no such ape-like creatures existed on the continent, it is hard to understand how they could have visualized such a design. Likewise, cave paintings from this era contain drawings of a large hairy, bipedal creature referred to as "Hairy Man." One such site is Painted Rock, located in the Sierra Nevada foothills of California. The pictographs found here include animals such as the coyote, bear, eagle, condor, frog, and lizard. Looming over these common animals is another more mysterious one: a towering, hairy two-legged creature that bares a striking resemblance to the modern-day Sasquatch. Since every other animal in the Painted Rock pictograph is a known creature, it is reasonable to assume that the Hairy Man is likewise a real creature, albeit a very large one with a shadowy history.

Real or not, by the 1700s the native tribes of Arkansas had more to worry about than hairy giants. As French colonial settlers began moving into the area, the Indians suddenly found themselves competing for their own land. At first they were able to coexist by establishing new social and political relationships, which benefited both peoples through trade and commerce, but this eventually collapsed when Arkansas came into the hands of the Union during Thomas Jefferson's Louisiana Purchase of 1803. To make way for American settlers, the United States concocted various treaties that ultimately forced the long-standing Arkansas tribes onto reservations far from their homeland.

When the population of Arkansas reached 60,000 in 1836, it was granted official statehood. The state's main contribution to the Union was farming, attesting to the rich, fertile lands found within its borders. As such, more and more settlers were drawn to Arkansas in search of opportunity. In conjunction with the growing commerce and population, newspapers began to spring up, documenting local happenings. In searching these documents, several early incidents of "wild man" sightings can be found.

The first incident was reported by both the *Arkansas Gazette* and *Memphis Enquirer* on May 9, 1851. In this account, two hunters near Greene County were startled when they came upon a very large, hair-covered animal as it was trying to catch a calf from a herd of cattle. When the creature noticed the men, it stopped its pursuit of the calf and just stood there eyeing them. Then suddenly, it turned and ran. They described it as "… an animal bearing the unmistakable likeness of humanity. He was of gigantic stature, the body being covered with hair and the head with long locks that fairly enveloped the neck and shoulders."

Upon investigation, the men reportedly found human-like tracks that measured 13 inches long. This is quite remarkable since the length correlates with the size of modern-day Bigfoot tracks. The reporter at the time theorized that the beast was in reality a human survivor of an earthquake that occurred in northeast Arkansas on December 16, 1811. But, of course, this hardly explains the gigantic height and the amount of hair covering the body. It sounds more like a modern report of a Sasquatch than a "wild man," but either way the theory does illustrate the tendency to classify these "monsters" as wild men during this time period.

Apparently, this was not the only time a "wild man" was seen in the area. The article goes on to state: "This singular creature has long been known traditionally in St. Francis Green and Poinsett counties. Arkansas sportsmen and hunters having described him so long as seventeen years since."

The second report dates to 1856 and describes a hairy "wild man" attacking a man in Sevier County north of Texarkana, Arkansas. The incident was first reported by the *Caddo Gazette* and was retold in the May 8 edition of the *Hornellsville Tribune*. This incident is often cited in popular books on the subject of Bigfoot and is considered to be more circumstantial evidence that Bigfoot-like creatures were living in the area as far back as newspapers began to report such sightings. In this account, a party of men were actually in pursuit of the wild man, which they described as " … a stout, athletic man, about six feet four inches in height, completely covered with hair of a brownish cast about four to six inches long.

He was well muscled, and ran up the bank with the fleetness of a deer."

The party was apparently trying to take the creature alive, so one of the men approached it on horseback. But the wild man did not find this to his liking: "... as the wild man saw the rider he rushed towards him, and in an instant dragged the hunter to the ground and tore him in a most dreadful manner, scratching out one of his eyes... and biting large pieces out of his shoulder and various parts of his body."

This incident is somewhat unique in that the creature acted violently. This is not the norm for the majority of Sasquatch encounters, but ironically, it does apply to the Fouke Monster, which has reportedly threatened humans at times. However, most Bigfoot sources fail to mention what occurred next, which would suggest that this may have been more of a wild *man* than a monster: "The monster then tore off the saddle and bridle from the horse and destroyed them, and holding the horse by the mane, broke a short piece of sapling, and mounting the animal, started at full speed across the plains."

The ability to commandeer a steed does not bode well as proof of an unidentified ape, although the line that states this was "an attempt to capture the famous wild man, who has been so often encountered on the borders of Arkansas and Northern Louisiana" does certainly cite a history of weirdness in the vicinity of Fouke.

Another interesting wild man incident can be traced back to 1865. In this fantastical account, a seven-foot "wild man" was captured in the Ouachita Mountains area near Saline County, Arkansas. Unfortunately the original source of the tale is unclear since the story was not widely publicized until it appeared in the 1941 book *Ozark Country* written by Otto Ernest Rayburn. Rayburn was a popular writer, magazine publisher, and collector of Arkansas/Ozark folklore who was commissioned by publishers to write the book as part of a series called American Folkways. The series, which includes a total of 28 books, was aimed at preserving the America's historical folkways such as Arkansas' Ozarks.

"According to Rayburn," write Janet and Colin Bord in the

1856: A hairy "wild man" attacks a man on horseback near the Arkansas/ Louisiana border.

Bigfoot Casebook: Updated, "the Giant of the Hills was often seen in the Arkansas Ouachita Mountains. This 7-foot *wild man* was covered with thick hair and lived in caves or by the Saline River. Everyone was afraid of him, although he does not appear to have harmed anyone. The decision was made to capture him, and the story tells that the men actually succeeded in this. They lassoed him in his cave and took him away to Benton jail. They also dressed him in clothes, which he tore off before escaping from the small wooden building. He was recaptured, but there the story suddenly ends."

Presumably this was word-of-mouth folklore known to Otto Rayburn at the time, or told to him during his research, but beyond that it is impossible to determine how well known the story was around Arkansas... or how truthful it is. As with many old-time folktales, the story may have been embellished or relevant parts of the story may have been lost so that what remains may be open to many interpretations. Since a final climatic ending is missing, it can probably be assumed that Rayburn did not completely fabricate the story and was only recording it as-is for posterity's sake. It would seem that if the story was in any way embellished by Rayburn, he would have certainly tacked on an appropriate ending about the beast's second escape and eventual return to the hills, or perhaps created a King Kong-esque twist in which he kidnaps a local woman and ascends to the top of the general store only to be shot down by the townsfolk. But sans such an ending, it would appear that the author merely called it as he heard it, adding credibility that this is a genuine tale. Whether the wild man in question is actually a Sasquatch-like creature, we may never know.

MANIMAL CONJECTURE

The mysterious "wild man" sightings are important to consider within the man-ape monster phenomenon for several reasons. First, they may suggest that humans are merely prone to seeing hairy creatures running amok in the woods. This could be due to an age-old instinctual fear of any large, hairy beast that might do us harm (just as we have a built-in fear of snakes), or perhaps they are reverberations of our own primitive self still ringing in our collective consciousness. Because there are sightings of man-apes throughout history, and in all parts of the world, it could be argued that it is something ingrained in our collective psyche. The fact that reports have always existed, while no definitive proof exists, underscores the theory that these beasts exist only in our minds.

Or perhaps the tales were made up and perpetuated to keep children out of dangerous areas. Those instances that exist only in folklore may suggest more of a fairy tale or warning type message. An article published in 2002 by the Arkansas Department of Parks and Tourism does a nice job of summarizing the origins of the state's most famous monsters and goes on to discuss possible reasons for their long history. The state boasts several alleged creatures, many of which fit into the ape-like category.

The article offers a possible explanation for the sightings:

> The fact that Arkansas was sparsely settled in the early 19th century helped create many of the original legends. Isolated and with virtually no social activities, families entertained themselves with folktales they had heard or ones they fabricated on the spot. Also, in an effort to keep children from dangerous bluffs, streams and caves, parents often invented frightening stories about those dangerous places.
>
> An example is at Sharp County's Cave City, where a cavern with an underground river is located. As the area was settled after the Civil War, the walk-in cave became a favorite rendezvous for youngsters until a story made the rounds that a "Mr. Jones" had entered the underground stream in a boat and never returned. This may have deterred some youthful explorers, but the cave remained a gathering place, even after the entrance was covered with steel bars in the 1930s.

Alternately, the wild man incidents could be attributed to ignorance of anomalies such as hypertrichosis, the medical condition that causes hair to grow all over the human body (a la wolf boys or girls of the freak show circuit). This is especially relevant to sightings of this kind that date back many years before such scientific explanations were widely known. Under these circumstances, it would have been natural to assume that these hair-covered humans were indeed *wild men* or *werewolves* of some sort; even more so, since any person suffering from such an aliment at the time was most likely shunned by society and forced to live in the woods. Thus, it is quite possible that these "wild men" were man-beasts of our own making.

It is also possible that some wild men were actually just that... feral humans... as was the case in 1875 when one was discovered near the town of Fourche in Central Arkansas. According to frightened witnesses, a "half wild animal" had been prowling around local farmhouses, stealing food, and forcing women to cook for "him" while the men were away at work. The culprit was eventually apprehended, and as it turns out, was actually a wild *man*. He had been employed by the Iron Mountain Railroad, but after suffering a mental breakdown he wandered off the job and took up residence in the woods for two years. A reporter at the time described the man as being the "wildest, greasiest, ugliest-looking, half-clad specimen of humanity it was ever our lot to behold." The man was subsequently tried by the local court after a judge declared him fit for trial. He was found guilty of vagrancy and sentenced to 60 days in a Little Rock jail.

However, occasional mistakes in identity or the existence of folktales do not successfully rule out the possibility that something far more mysterious is responsible for the sightings. The reality may, in fact, be more straightforward — or perhaps more frightening — in that the historical reports of wild men are the equivalent of modern day Sasquatch reports. Yes, just possibly, hairy man-ape creatures existed in the American backwoods long before we ever coined names such as Bigfoot, Fouke Monster, or any of the dozens of indigenous names that have been applied to them. In those days the moniker of "wild man" was as fitting as any and that's what was reported or told to neighbors in an effort to describe the beast.

But following the reports of the 1800s, the descriptions would move away from terms such as "wild man" and take on a decidedly more "ape-like" tone. Perhaps the public's growing familiarity with more worldly animals would help in their efforts to describe the hairy, shadowy creatures they began to find as they settled the area.

Since these formative years, Arkansas has gone on to contribute a great deal to American culture. The state has contributed to folk, country, and bluegrass music. It has given rise to champions of civil rights and been the humble birthplace of legendary entertainers such as Johnny Cash and Billy Bob Thornton. It has even given us a U.S. president. So it is not surprising that among all the honors and scars, the state has also managed to contribute a lasting entry to the canon of cryptozoology creatures with a certain Deep South flair. Just as Jersey has its Devil, Missouri has its Momo, Florida has its Skunk Ape, and so on, the Fouke Monster is a true creature of his environment. Crawling up from the swamps of the Sulphur River and traveling the meandering paths of Boggy Creek on hot sweltering nights, the monster reeks of Southern swagger, occasionally letting out a bellow from the dark corner of the forested backwoods. There's no well-groomed, well-fed California Bigfoot here. No, the beast of Fouke is lithe, lean, and covered in long matted fur. And he's got a knack for frightening the locals. Yes, the world would eventually find out what the quiet residents of the area had suspected all along: that they were sharing the land with a strange and unknown animal.

2. FOUKE LORE

BIRTH OF A MONSTER

A creature resembling what came to be known as the Fouke Monster first came to the attention of local officials in 1946, although it wasn't widely reported at the time. Not until the mid-1960s did encounters with some sort of peculiar animal began to occur with any frequency around the Boggy Creek area, but even these did not receive widespread attention. During that era, the term "Fouke Monster" was not yet used to identify the animal among the locals. It wasn't until the early 1970s that the small town called Fouke became the nationwide focal point and namesake for the hair-covered beast now referred to as the Fouke Monster, or sometimes as the "Boggy Creek Monster." It was then that the monster gained real popularity and sank its claws, so to speak, into the media. Newspapers such as the *Arkansas Gazette*, *Texarkana Gazette*, *Texarkana Daily News*, and Little Rock's *Dispatch* began to run stories on the various reports. These were eventually picked up by the Associated Press and United Press International and transmitted to newspapers across the nation, drawing an unprecedented level of attention to the small town.

The first of these reports, considered to be the starting point for the monster's media age, appeared in the May 3, 1971 edition of the *Texarkana Gazette*. The article, written by journalist Jim Powell, told of a hair-raising experience reported by the Ford and Taylor families, both of whom had recently moved to the north side of Fouke about five miles from the infamous Boggy Creek. The climactic incident, which followed several consecutive nights of strange happenings, took place after nightfall on Saturday, May 1.

FOUKE FAMILY TERRORIZED BY HAIRY MONSTER
"I was moving so fast I didn't stop to open the door,
I just ran through it."

This was the statement of Bobby Ford, 25, of Rt.
1, Box 220, Texarkana, Ark., Sunday morning after an
unidentified "creature" attacked a house about 10 miles
south of Texarkana on U.S. Highway 71.

Ford was taken to St. Michael Hospital where he was
treated for scratches and mild shock and released.

"We have lived here only five days and I think we are
going to move now," Patricia Ford, 22, Bobby's sister-in-
law said. "The thing has been to the house three times
now."

The "creature" was described by Ford as being about
seven feet tall and about three feet wide across the
chest. "At first I thought it was a bear but it runs upright
and moves real fast. It is covered with hair," he said.

Ford, his brother Don Ford and Charles Taylor saw
the creature several times shortly after midnight Saturday
and shot at it seven times with a shotgun.

The incident was utterly bizarre and frightening to the families,
but apparently this wasn't the first time the creature had been
lurking around the house. Earlier that week the wives had heard
something walking on the porch. On Friday night it tried to break
into the house.

The article continued:

Elizabeth Ford said she was sleeping in the front
room of the frame house when, "I saw the curtain
moving on the front window and a hand sticking through
the window. At first I thought it was a bear's paw but
it didn't look like that. It had heavy hair all over it and
it had claws. I could see its eyes. They looked like coals
of fire ... real red," she said. "It didn't make any noise.
Except you could hear it breathing."

Ford said they spotted the creature in back of the
house with the aid of a flashlight. "We shot several times
at it then and then called Ernest Walraven, constable of
Fouke. He brought us another shotgun and a stronger

light. We waited on the porch and then saw the thing closer to the house. We shot again and thought we saw it fall. Bobby, Charles and myself started walking to where we saw it fall," he said.

About that time, according to Don Ford, they heard the women in the house screaming and Bobby went back.

"I was walking the rungs of a ladder to get up on the porch when the thing grabbed me. I felt a hairy arm come over my shoulder and the next thing I knew we were on the ground. The only thing I could think about was to get out of there. The thing was breathing real hard and his eyes were about the size of a half dollar and real red.

"I finally broke away and ran around the house and through the front door. I don't know where he went," Bobby Ford said.

At that point, everyone in the house saw it move rapidly into the field next to the house and then it was gone. Constable Walraven was called back to the scene at 12:35 a.m. where he promptly searched the area. He did not find any trace of the strange creature. It had simply vanished into the dark woods behind the house.

With the coming of daylight, the men could investigate more thoroughly. They found pieces of metal around the bottom of the house that had been "ripped away." They also noted window damage and scratches on the front porch. In the soil around the house, they discovered curious tracks, which may have been left by the creature. Whatever it was appeared to have only three toes, which was puzzling since this is not characteristic of any known bipedal creatures.

The following day, May 4, a reprise of the story was printed in the *Arkansas Gazette*. The article read, in part:

Bobby Ford, 25, who said he was attacked by a "large, hairy creature" at his home was treated at a hospital here early Sunday for scratches and shock.

Ford, of Fouke, a small Miller County community 15 miles southeast of here told county officials and Fouke

Constable Ernest Walraven that about midnight Saturday a creature poked its paw through a hole in a window.

Ford said he and three other adults chased what they described as a large, hairy animal into a wooded area behind the house.

Later, Ford said something kicked in the back door and they again saw the creature behind the house. As he was returning to his house after chasing the creature, he said the animal knocked him down. Ford said he escaped and ran into the house.

Walraven and county officials searched the area Sunday and said they found several large tracks.

It is important to include the complete text of these articles rather than merely summarize the story, since it more accurately conveys the flat, matter-of-fact tone in which it was reported at the time. Devoid of modern-day cynicism that usually accompanies such "creature feature news" today—whether in actual newsprint or on television or internet—it is easy to imagine how this struck a chord of fear within the quiet melody of Fouke and the surrounding areas. Not only did Mr. Ford believe he saw a monster, he ended up in the hospital to be treated for injuries and shock!

And this was only the first of such horrifying reports.

The second incident was made public in the May 24, 1971, edition of the *Texarkana Daily News*. This time "a large hairy creature" was seen crossing a road several miles south of Fouke near Boggy Creek, as witnessed by Mr. and Mrs. D.C. Woods Jr., and Mrs. R.H. Sedgass, all citizens of nearby Texarkana. As the group was traveling north on U.S. Highway 71, just seven miles from the Ford house, they saw a thing with "dark long hair" run upright across the road in front of the car and disappear into the darkness beyond. Mr. Woods first thought they were going to collide with the animal, but due to its unusually fast pace, it ran past unharmed. The variety of eerie descriptions given by the witnesses indicated that "it was swinging its arms" as it ran and "looked like a giant monkey." The group had recently read about the Ford incident, so they were all aware of the monster at the time and even stated that they "thought it was just a hoax." But after what they saw that night on the lonely stretch of Highway 71, their minds were changed, as was very clear in the article. Mrs. Sedgass was quoted as saying: "Some people don't think there is anything to it (the monster), but I do," summing up the feeling that the trio shared.

The Fords may have been newcomers to the area, but the Woods were well known and respected citizens. While many of the locals were suspicious of the Fords' tale, they were hard pressed to explain away the Woods' sighting. The Miller County Sheriff at the time, Leslie Greer, was quoted as saying, "I know those people, and they were very reliable and very truthful. I don't know what they saw, but I do believe they saw something."

Both of these original monster articles were written by Jim Powell, who could not have imagined he would become a major player in a legend. At the time he was a hard working journalist who was simply covering a bizarre story. But his diligent effort and straightforward presentation of the incidents were just the right combination to spark major public interest in the "monster."

"I just wrote what people told me," Powell explained when I spoke to him on the phone. Powell still resides in Texarkana. "They didn't editorialize like they do now. They just printed the story."

The morning after the Fords had their terrifying encounter, Powell got a call from Dave Hall, news director for Texarkana's KTFS radio station. I also spoke with Hall during my research. He gave me a rundown of how it all started: "A doctor friend of mine called that morning [May 2] and said he had a guy down there at the hospital that had been attacked by a monster. He was scared so bad he was in shock. So I called Jim and said 'let's go down there and see what this is all about.'"

They wasted no time in getting an address for the incident and hurried to the scene. When the two men arrived, they found the Fords in a state of frenzy, packing their belongings into a U-Haul in a tremendous hurry to leave town. It was certainly odd since the family had only moved there less than a week earlier.

Several area law enforcement officials were on the scene along with a throng of onlookers that was growing by the minute. Powell retraced the events of the previous night, trying to sort out what

The actual house, pictured in 1969, where the Ford incident occurred.
The Fords rented the property from the Simmons family.
(Courtesy of the Miller County Historical Society)

had happened. In our phone conversation, Powell told me that "[Bobby Ford] definitely thought he was being attacked. He fought with something, then ran around the porch, and nearly jumped right through the front door. He was really afraid."

Powell and Hall searched the immediate area for evidence. There was a freshly plowed field behind the house, so the men looked for tracks where the Fords said they had seen some glowing eyes. "We went into the area behind the house and saw unusual footprints, and small saplings broken off," Hall told me. "We never saw any blood, although the people said they fired several shots and thought they hit it."

The *Texarkana Daily News* and *Texarkana Gazette* both published a follow-up article in the days following, theorizing that the mysterious visitor may have been something less than monstrous. The article, which includes Powell's first reference to the "Fouke Monster" by name, shifts the blame to a wild cat. Its headline read: "Monster may be mountain lion" and included the following statement: "'We think now it might have been a big cat, like a mountain lion or puma,' Don Ford said Monday while sitting on his porch watching people wander through his fields looking for a trace of the 'Fouke Monster.'"

Rumors also circulated that the perpetrator had actually been a horse. Several sources agreed that an old horse had been seen lumbering around the area of Highway 71 where the Ford house was located. It often trampled through gardens and approached houses at night. Some claim that the old horse was found dead a short distance from the Ford house the following day, having been killed by a shotgun blast. However, other sources deny that such a horse was ever found. Either way, it does seem odd that no hoofprints were noted around the Ford house, nor was a blood trail ever found.

Despite the new theories, the original Ford tale had ignited a sense of public curiosity that could not be easily doused. The story had already gone viral, and it would become clear that no efforts to debunk or calm the situation would be entertained. After all, how could a four-legged cat or rickety old horse be mistaken for a seven-foot upright monster? A legend had been born and nothing would stop it from growing.

IT WALKS AMONG US

Two days after the Woods saw a strange creature lope across the highway, the *Texarkana Gazette* printed another small article, indicative of the burgeoning interest in the Fouke Monster. The short report, featuring the rather academic headline of "State Funds Sought for 'Monster' Hunt," put forth one citizen's suggestion that state officials should take interest:

> Dean Combs, a history instructor at Dial Junior High School in Pine Bluff, Ark., is taking the Fouke "monster" seriously.
>
> Tuesday Combs said he believes the "monster" is a symbol of Arkansas' wild heritage and that the state should appropriate money for the capture and preservation of the beast.

State officials were probably either scratching their heads at such a notion, outright laughing, or just hoping it would go away as quickly as it started. But as the reports continued to roll in, it was becoming apparent to everyone that something unique and exciting was happening down in Miller County. Folks didn't have to wait long for another titillating report to hit the pages of the *Texarkana Gazette*.

It was June 2, only a month since the first incident sparked the creature to life in the minds of the locals. Officials responded to a call from three individuals claiming that they had seen a "tall hairy creature with red eyes" during the evening hours in a wooded area near Texarkana. According to one of the witnesses, Gloria Dean Richey, they spotted the creature "squatting" on an embankment across the street from her Oats Street residence after hearing something walk through a weeded area near the road. "We shined a flashlight on the spot and saw the creature," Richey told reporters. "He was real tall and hairy and had real red eyes." When the light hit the creature, the dogs began to bark wildly. At that point the thing started running through the heavy brush "leaping high over the weeds and running faster than a man could."

Mrs. Richey glimpsed the animal once more, approximately 25 yards from where it first entered the heavy brush, before it finally disappeared for good. At that point, the two other witnesses, Junior Goodman and Jerry Smallwood, went to retrieve their guns and call police. When they returned, they followed a trail of trampled weeds and bushes, but could not find any further trace of the creature. Mrs. Richey waited in the house for the men to return. Her fright was evident: "I have never seen anything like it. I know it wasn't a cat or a man. I could still hear the dogs acting up and the brush

June 2, 1971: The creature is seen in the vicinity of Oats Street north of Fouke.

breaking and rattling. I didn't go out again until [Goodman and Smallwood] got back."

The police searched the wooded area around the house shortly after receiving the call but came up empty handed. It was obvious that something had scared the trio, but it was impossible to substantiate their story without tracks or some other evidence. To make things even more puzzling, the location was much further from the usual Boggy Creek haunt. The first incident—at the Fords—occurred in the vicinity of the north fork of Boggy Creek (a.k.a. Chicken Creek), while the second incident—the Woods— occurred on Highway 71 near the south fork of the creek. Oats Street is located on the Arkansas side of Texarkana, about 10 miles north of Fouke. In those days it backed up to an area of thick trees, but still, it was much closer to the outskirts of a large city than any of the previous sightings.

However, it was hard to completely dismiss the possibility that the "monster" had decided to venture north from the creek. The following Saturday, June 5, police received yet another call from a residence at the intersection of Oats and Washington. This time a child had reported seeing a "monster in the woods across from a group of houses in the area." As unreliable as a child's eyewitness "monster" account may be, it still added fuel to the fire and caused some alarm in the neighborhood.

A more thorough search of the area, at the request of the residents, turned up some unidentified tracks at an abandoned fertilizer plant on North Oats Road. On Sunday June 6, Miller County Constable Paul Jewell and Deputy Constable Richard Haygood discovered what they described as "four-inch wide tracks" in the soft soil surrounding a large barn-like structure used for storing fertilizer bins. There were at least five visible prints, but unfortunately they could not tell if the animal's foot had claws because of the nature of the soil. It seemed that for the time being the creature had managed to cover its tracks.

But not for long.

On the hot morning of Sunday, June 13, Yother Kennedy discovered a series of mysterious footprints in his freshly plowed soybean field located near the south fork of Boggy Creek. The tracks originated from the woods at one corner of the field and traveled about 150 yards before disappearing into the trees on the other side. The trackway appeared to have been made by a bipedal creature walking upright. According to an article that ran two days later in the *Texarkana Gazette*, the tracks measured 13.5 inches long by 4.5 inches wide with a maximum stride of 57 inches between them. Just as in the Ford incident, the animal appeared to have three toes, all about the same length. Another smaller toe imprint was observed about five inches back from the big toe, but this digit only made a faint indention in the sandy soil.

A curious incident that occurred just prior to the track find was also reported by Kennedy: "Kennedy said he was plowing Wednesday and had stopped to work on his tractor when he heard strange noises coming from the thick undergrowth near the field. He said he got his rifle and plowed the rest of the day with it close by. Then Sunday he returned to the field to see how his beans were doing and found the tracks."

Several Fouke officials and citizens were initially called to the scene that Sunday morning, including Constable Ernest Walraven, Sheriff H.L. Phillips, J.E. "Smokey" Crabtree, and Willie Smith, who owned the land. The tracks were so unusual that even doubters began to wonder what was going on. Walraven, who had previously investigated the Ford incident, seemed swayed by the evidence. He told reporters: "At first I didn't think too much of the sightings but now I do. I have never seen tracks like this and I have been in the woods all my life."

Sheriff Leslie Greer and the local game warden, Carl Gaylon, also investigated. Neither had ever seen animal tracks like those before, so they could not make a judgment as to whether they were authentic, only that they were indeed mysterious.

Willie Smith weighed in on the discussion, informing the press that he and his family had seen these type of tracks in the area many times in the past. He was also the first to theorize that the creature must live and roam in the vicinity of the creeks, which more or less spread out in a ring around Fouke. "Every time it has been seen around here, it has always been near one of the creeks," he told reporters.

Smith also cited numerous broken limbs in the area that he believed were indications of a tall animal walking through the trees. To strengthen the case, Smith's niece-in-law[3], Bobbie, told of her own sighting, which had occurred just two weeks earlier. She said the creature stood approximately six feet tall as she watched it walk through the woods near the soybean field at about 7:30 p.m. one evening.

Word of the track find spread quickly, and more of the locals came down to the field to see for themselves. Rick Roberts, whose father served as mayor from 1978-1991, was one such person. At the time he was a young man, familiar with the Sulphur River Bottoms and the surrounding area. He took a special interest in the phenomenon unfolding before him and would later own the local Monster Mart convenience store in Fouke. It was his mother, Jane Roberts, who was responsible for helping law officials make plaster casts of the footprints that day. Fouke officials had very little experience in casting at the time, so Robert's mother, skilled with arts and crafts, assisted them. Like the others, Roberts didn't know what to make of the tracks, but he was definitely impressed by the distance between them. In a series of personal interviews I had with Roberts, he told me: "If the tracks were a hoax, they would have been very hard to fake. It would not have been easy for a person to get that much distance between each one."

Journalist Jim Powell and radio personality Dave Hall represented the press that day. They had both been present at the Ford investigation, so naturally they were interested by the new evidence this case presented. But Powell was not as convinced as Walraven that the tracks were those of a real animal or even an unknown creature. "I noticed that it stepped over the plants," he told me during our phone conversation. "I've never seen an animal that didn't step on plants as it crossed a field. It just didn't seem right." This was puzzling. Those present agreed that whoever had made the tracks had been reluctant to step on any of the young bean plants.

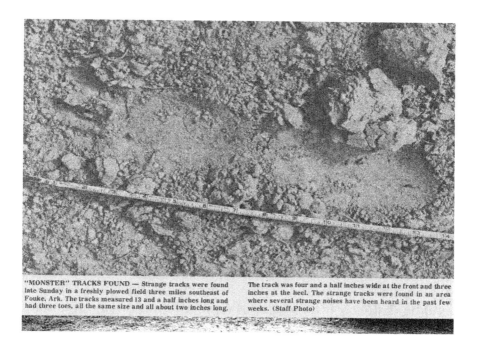

"MONSTER" TRACKS FOUND — Strange tracks were found late Sunday in a freshly plowed field three miles southeast of Fouke, Ark. The tracks measured 13 and a half inches long and had three toes, all the same size and all about two inches long. The track was four and a half inches wide at the front and three inches at the heel. The strange tracks were found in an area where several strange noises have been heard in the past few weeks. (Staff Photo)

One of the tracks found in Willie Smith's soybean field.
(Courtesy of the Texarkana Gazette)

Regardless, there was no denying that *something* had made strange tracks in the soil, whether it was a strange animal or just a clever hoaxer. Since there was no solid evidence pointing either way, it was simply news and Powell reported it that way. The resulting story appeared in the June 15 edition of the *Texarkana Gazette*, taking up the better part of a page that featured a large photo of a single track and the headline: "Monster Tracks Found."

By now the Fouke Monster was big news, and his story was being followed by a good portion of readers in Texarkana and the surrounding areas. The lawmen were taking the reports seriously and were doing everything they could to solve the mystery. But so far, every trail went cold. The monster might have left tracks for all to see, but he himself would not be so easy to find.

MEDIA MAYHEM

It wasn't long before other regional newspapers got wind of the strange discovery and ran stories in their own editions. Photos of the bizarre three-toed tracks, along with more revelations offered up by the landowner Willie Smith, spread to a wider audience. These new anecdotes had to do with encounters he and his family had with the monster years earlier, giving even more credence to the belief that something very strange was lurking in the spooky woods along Boggy Creek.

In an article printed in the *Victoria Advocate*, Smith claimed that his sister had first seen the creature way back around 1908 when she was just 10 years old. This pre-dated any previously reported sighting of the creature by almost 40 years. Smith went on to state that he himself had first seen the creature in 1955 near his house along Boggy Creek. "I thought he was a man. I shot at him 15 times with an Army rifle, but missed," he told reporters.

Despite the clamor of gunshots that the mystery creature experienced, it did not stop him from returning for an encore. Smith went on to add: "Next time he came up behind the house throwing chunks at my dog. So, I shot through the brush and missed him again."

Fortunately for the manimal, Smith was either drunk or the worst shot in Fouke.

The next incident hit the papers on June 16, 1971. In the early morning hours, two local residents witnessed a creature, which fit the monster's general description, "slouch" across a gravel road in front of their car. The road ran two miles south of Fouke, only a quarter mile from Smith's bean field. Al Williams and A. L. Tipton, both residents of the rural community, reported the sighting to Sheriff Greer, saying that they were close enough to see that the creature was either a "small ape or large monkey." Tipton stated that "it appeared to be about three or four feet tall as it crouched over and walked across the road." Although the height estimation seems

Mayor Virgil Roberts holds an original 1971 track casting.
(Courtesy of Rick Roberts)

at odds with the usual seven-foot range, it did reinforce the theory of it being ape-like. If it was a real animal, then it likely had offspring, which could explain this creature's smaller stature. Theories of all kinds abounded.

Another *Texarkana Gazette* report from late June told how two men from Kansas stopped into town to ask about what kind of wild animals were thought to inhabit the area. Several of the Fouke locals figured the two men were joking about "the monster" and immediately laughed it off. But when the locals began to speak of their would-be beast, the men were shocked. They claimed to have no knowledge of the recent incidents. They were only concerned because they had seen some kind of peculiar looking two-legged animal standing by the side of the road. They may or may not have been pulling a fast one, but they were indeed from Kansas. One of the locals saw their car's license plate.

Willie Smith holding a trophy deer outside of
the Boggy Creek Café, circa 1975.
(Courtesy of the Miller County Historical Society)

Another alleged sighting by outsiders was not reported to the news at the time. Only later did Sheriff Greer reveal that a group of several women and children, who had traveled to the area to look at the tracks in the soybean field, reported seeing an ape-like creature nearby.

The monster also inspired some pranks, as such things tend to do. On June 28, three local Fouke men claimed that they were attacked by the creature, showing claw marks as evidence. But when Sheriff Greer noticed traces of blood under their fingernails, the truth came out that the men had simply been drunk and got into a fight amongst themselves. He threatened to arrest them but instead fined them $59 each for filing the bogus monster report.

In addition to the regional newspapers, some of the stories were also picked up by the associated news services, which allowed them to be reprinted in papers around the country. Outside reporters began to tie in other Arkansas strangeness as they covered the subject. One subject they brought up was the legendary White River Monster of the Ozark Mountains. Described as a 30-40 foot water serpent, this would be the equivalent of the Loch Ness Monster living in northern Arkansas. Though obviously more of a stretch on any believability scale, it did nonetheless serve to heighten the monster mania.

By this time, the Fouke Monster was basking in the limelight of notoriety, capturing the public's attention with a steady stream of tales chock full of fright and intrigue. As more and more people read the accounts, and in some cases began to drop by the area to have a look, the creature was well on his way to becoming Fouke's own contribution to Deep South lore, forever linked to the town in name and legend. The monster had seemingly been lurking in the Sulphur River Bottoms for most of a century, but for Fouke...this was only the beginning.

BOUNTY ON THE BEAST

Shortly after the Ford incident hit the papers, KAAY radio in Little Rock, Arkansas, offered a $1090 reward for the monster's capture. Station representatives, who felt they were doing a public service by helping Fouke rid itself of its "monster problem," announced by way of the *Texarkana Gazette* that the reward would be paid to anyone who handed over a "legitimate and valid monster" to an authorized representative of the station. Further rules required that the monster be "alive and in good health at the time of delivery," and that no property could be damaged in the process of capturing said creature. One final stipulation stated that: "All monsters turned in will become the property of the station." This, however, was contrary to a statement made by Sheriff Leslie Greer, who said that "if the culprit is captured it would become property of Miller County and not that of the radio station."

To up the ante, a local man by the name of Raymond Scoggins offered up his own $200 bounty. Like the radio station, he stipulated that the creature must be brought in alive. Scoggins, who had lived in Fouke before moving to Texarkana, had heard reports about a mysterious monster for nearly a decade before the recent rash of sightings. He believed it to be "a member of the ape family." To make it official, three Fouke residents were appointed to verify authenticity of any captured beast: Mayor James D. Larey, City Marshall Bob Bowen, and Constable Earnest Walraven.

The posting of these bounties sparked an all-out hunt and signaled the beginning of what would be several years of Fouke Monster frenzy. Monster hunters began to descend on the little town, looking for a chance to bag some truly exotic game. The radio station was contacted by groups from nearby cities, including Benton, Pine Bluff, Conway, Texarkana, and Little Rock, as they announced plans to send search parties toward Fouke.

This quickly became more of a concern for town officials than the actual monster, as trigger-happy hunters began to run amok in search of the beast. As well, calls began to jam up official phone lines and letters started showing up at the mayor's office. Even reporter Jim Powell was getting calls and letters inquiring about the monster and the ensuing hunt.

The Miller County Sheriff's Office first tried to control the situation by stopping people to check for guns and liquor. Trespassing was a huge problem, so visitors were asked not to cut any fences, although they did anyway. A long time friend of mine, Larry Moses, remembers when his father headed up to Fouke to join in the hunt. "He grew up in east Texas and every liquor-crazed teen from around Texarkana went to try and track it down," Moses told me. "I think it just ended up as an excuse to go out, get drunk, and try to shoot something."

At one time or another an estimated 500 hunters stomped through the woods around Fouke looking for the monster. Eventually, officials had to flat-out ban the use of firearms unless it was actual hunting season, in an attempt to avoid a fatal accident that seemed to grow more likely as the bounty hunters continued the hunt. Looking back now, it's amazing that no one was hurt. "It was chaos," Rick Roberts told me, as he recalled the craziness.

"There were a lot of people out there looking who weren't local."

During my research I also spoke to H.L. Phillips, who was a deputy in Miller County at the time.[4] He remembers the situation well. He told me that the official stance of local law enforcement was to take the "monster" reports seriously, but at the same time try to downplay the incidents in an attempt to minimize the madness. "A lot of the stuff that was reported to us never made it to the media," he told me over the phone one afternoon. "We tried to downplay it because every time something would come out, the place would be loaded down with people with guns wanting to go in and try to find it."

Along with the hunters and their guns, people showed up with tape recorders and cameras trying to capture the monster's scream on tape or snap a photo to finally prove its existence. Perhaps not surprisingly, with all the fuss in the forest, the monster stayed out of sight, effectively avoiding the hunters but also casting doubts as to whether there was really anything to the stories.

During this first wave of monster mayhem, only a few enterprising locals attempted to make good from it. Willie Smith, who owned the Boggy Creek Café at the time, created casts of the three-toed prints found in his bean field and sold them to monster enthusiasts eager to purchase any kind of souvenir during their visit. To increase the appeal, the casts were autographed by both Smith and Smokey Crabtree. Since there were not many places to stop for a bite around Fouke, the Boggy Creek Café was a perfect outlet for Smith's enterprise.

One particularly industrious young man, Perry Simmons, who was 15 at the time, made himself a tidy profit by giving tours in and around the house where the Ford incident took place. Just after the report hit the papers, people began to line up in cars along the street trying to get a look at the house, which was owned by his stepfather, Joe Simmons. As will happen when sightseers drop by unannounced, the property was being overrun, and in some cases people were entering the house without permission. To keep people out, a barbed wire fence had to be installed, but that only made it even more of a spectacle. At times, there were an estimated one hundred people gathered on the street to see the house for themselves.

Recognizing a perfect business opportunity, Perry posted a sign that announced "Guided Tours for $1" and in no time he was taking in money to give curious folks a peek in the house and a look around the field behind it. It became something of a residential carnival attraction. People offered as much as $5 in their excitement, as if they were going to actually get a glimpse of the monster itself. In fact, people were so overcome by monster fever that when passing through the vegetable garden in back of the house, they noticed trampled corn stalks and believed that the monster must have surely been the culprit. Perry simply smiled, knowing that in reality raccoons had done the damage. But it just didn't seem right to spoil the fun. After all, people were paying good money for a close-up look at the mystery.

Of all the events, the Ford incident became the cornerstone of the monster craze. Not only had it been the first of the reports in the 1970s, but it had the most substance to it: prolonged monster scuffles, a hairy paw, gunfire, and a trip to the hospital. Despite alternative explanations proposed by news reporters, there was still no refuting the fact that something strange went down at the house. And as far as the public was concerned, a monster was the best explanation. The subsequent sightings only served to back up this conclusion.

Law officials were not prone to dismiss the whole affair either, and for good reason: one of their own had seen it with his own eyes! According to H.L. Phillips, a Deputy Sheriff by the name of Robinson nearly hit a creature fitting the description of the Fouke Monster as he was patrolling in his car one night. In the official report filed with the Miller County Sheriff's Office, he claimed that he could see the thing very clearly as it took two or three big leaps across the road in front of him, jumped over a fence, and disappeared into the woods. In the opinion of Phillips, he was a good officer with solid judgment, not someone prone to wild stories.

In general, however, the townsfolk of Fouke were dubious of the whole affair. They were at the center, but the limelight was

1971: The creature is spotted by a law enforcement official as it ran in front of his patrol car.

*Boggy Creek Café / Smith's Service Station sign circa 1970s.
(Courtesy of the Miller County Historical Society)*

not something they were used to or even liked. The town's mindset was to minimize the craziness; they didn't want to get caught up in the trappings of tourism. They were far more concerned about upholding the peace, reducing property damage, and staving off any causalities that could result from the gonzo hunts. It would be some time before the town came to realize they had become synonymous with a cryptozoological icon.

Don't Blink

When reporter Jim Powell referred to the alleged creature as "The Fouke Monster," it was apparent early on that the name was going to stick. It is certain that the town of Fouke did not intend to have a monster associated with its name, but as history has proven, it is a pairing that will never be broken as long as the stories still propagate on the internet and in documentaries, and authors continue to write books about such creatures. Although it is sometimes referred to as the "Boggy Creek Monster," due to its movie appearances, it is not long into a conversation before "Fouke" is attached to its name. So, to fully understand the monster, it is important to get an accurate picture of the small town that lies at the heart of this story.

As mentioned previously, Fouke is part of Miller County which shares borders with both Texas and Louisiana. Despite the popular dry, flat image of Texas, the eastern areas of the Lone Star state are quite heavily thicketed, as are the great expanses of northern Louisiana, which all blend together as the trio of states merge with the rich countryside of Arkansas. It is here that Fouke occupies an unassuming position among the majestic pines on a scenic stretch of old U.S. Highway 71, also known as Monster Expressway until recently (yes, this was an official name!). A few minutes northwest of Fouke lies the major city of Texarkana, named in honor of the three surrounding states (Tex-Ark-Ana). To its south lies the Sulphur River Bottoms and the point where the Sulphur and Red Rivers converge.

At the time when the monster was seen crossing U.S. 71 in front of Mr. and Mrs. D.C. Woods, Jr., it was the only highway that went anywhere near Fouke. Since then, the more modern Highway 549 has been added. Opened in 2004, U.S. 549 originates in Texarkana and runs parallel to U.S. 71 as it heads toward Louisiana. It adds convenience, but to pass through Fouke's main strip, you would still need to traverse the old Monster Expressway.

While Fouke is considered part of the ever-expanding Texarkana Metropolitan Area, it still manages to maintain a rural, down-home feel even to this day. That is not to say that Fouke has remained behind the times—it has plenty of modern conveniences—only that it maintains a feel of traditional country culture that one might think of when seeking a truly out-of-the-way retreat or perhaps a place to begin a hunting trip. Even by small town standards, the town of Fouke is small, with a mere 852 people making up its total population, according to recent real estate reports. At the high point of its monster mania back in the early 1970s, the population was barely over 500, which illustrates its minimal growth over the last 40 years.

When leaving the main highways of the Texarkana loop and proceeding down U.S. 71, the road quickly sheds the landmarks of modern life and takes on a more peaceful demeanor. If you've ever had the pleasure of taking in one of the vast rural areas of Arkansas, then you know the rich green beauty that rises high on either side of its roadways and seems to swallow the spattering of houses and other buildings, which dare to share the land. The trees here are thick, as if they are attempting to keep in their secrets, and the brush threatens to engulf anything that is beyond the reaches of a powerful lawnmower. The occasional skeletons of rusted farm equipment and cars only add proof to the constant battle between man and nature that exists outside the cement confines of the big city.

As you proceed through Fouke, only a small hint of the town's dubious monster can be seen in the form of a convenience store called Monster Mart. Over the years, the establishment has been a steady reminder of the small town's big legend, proudly displaying an artist's rendering of the creature on its outer wall and offering a small display and curio counter inside where Fouke Monster gifts can be purchased along with the usual assortment of traveler's treats and guide maps.

But don't let a preconceived idea of rampant capitalism influence the vision you may be conjuring of the Monster Mart's offerings. There are no large displays of t-shirt styles, coffee mugs, plastic figurines, or anything that might be associated with a big-city tourist operation. The Fouke Monster merchandising is limited in scope, offering one shirt design with a crudely drawn image, a small assortment of postcards, and a few other items displayed behind a small glass counter. While it does contain one plaster cast of an alleged Bigfoot track, this is not one of the curious three-toed tracks found in the bean field back in 1971.

Tacked to a small bulletin board near the entrance are some newspaper clippings, but these don't seem to be given much respect considering they are not laminated or even arranged in an orderly fashion. The clippings are yellowed, or worse, just copies that flutter like frantic moth wings every time someone walks through the door. Given the sparse amount of monster memorabilia, it is evident that either the Fouke Monster doesn't demand such money-luring fanfare these days or else the Monster Mart shares the town's general reluctance to fully embrace its notorious resident.[5]

Next to the Monster Mart, in a small lot between it and the Fouke Post Office, tourists can find the only other visible reminder of the town's legacy. This is a large version of the monster made out of metal, artistically painted to look like fur. The head has an oval cutout where monster hunters of all ages can insert their own grinning face in order to capture the moment with a photo. A plaque hangs in front by a chain, identifying it as the "Boggy Creek Monster - Fouke, AR."

On my first trip to Fouke, I spoke with former Monster Mart owner Rick Roberts, who graciously entertained me for more than two hours with stories of the monster and its rise to cult movie infamy. Roberts assured me that tourists still stop by the store on a daily basis to ask about the monster or to purchase t-shirts or postcards. We spoke about an episode of the ABC television show, *Wife Swap*, which had ironically aired the night before I made one of my many trips to Fouke for research. In the episode, one of the families told how they spent their leisure time "monster hunting" in a fun-loving, amateur cryptozoologist sort of way. As part of the show, the wife took the children of the swap family on a monster hunt to show them how fun it can be to get out in nature. So where did they go? Fouke, Arkansas. Even though the narrator kept pronouncing it *Fook*—I could almost see the eyes of the entire town rolling every time he said it—this underscored the fact that Fouke Monster charm was alive and well, no doubt about it.

Leaving the Monster Mart and continuing south on U.S. 71, it is only a scant few rotations of the car tires before the main strip of Fouke is already in the rearview. It's usually at that point you begin wonder where in the heck is the *legendary* Boggy Creek? That was my question the first time I visited Fouke. Returning to the Monster

Mart, I was told that the infamous Boggy Creek was about two miles on down 71, but don't expect so much as a sign marking its location. And indeed, Boggy Creek, for all its big-time Hollywood fame and monster trafficking, is but a thin tributary of water that snakes through the trees and runs under U.S. 71 in quiet, dignified solitude. In fact, if you didn't know it was there, you would drive right past it on your way out of town.

Most all bodies of water, no matter how insignificant, usually get a state-issued green sign at the point where they cross a highway, or at least that's what I've noticed in the Texas-Arkansas area. But this is not the case for Boggy Creek, which seems an injustice considering its larger-than-life reputation. I later asked Rick Roberts why there

Metal monster standing near the Fouke Post Office.
(Photo by the author)

42

was no sign, speculating that perhaps it was because it was being stolen as a memento too often. He confirmed my suspicion, but indicated that moreso it was because of safety. Too many people were stopping on the roadside, creating a hazard.

Despite the absence of the state name tag, the notorious waterway is not totally devoid of recognition. After a few visits, I was clued in on an inconspicuous plaque that is set into the concrete at one side of the bridge. There, molded in heavy brass, are the words "Boggy Creek 1970." So, even though it is not evident to the casual motorist, the little creek is not completely condemned to anonymity. In a way, it seems to symbolize the town's love-hate relationship with its monstrous fame. At first glance, there may only be a few reminders of the monster's presence, but there is certainly more hidden beneath the surface.

Rolling out past Boggy Creek, U.S. 71 is again engulfed by the rich treeline on either side of the road, making Fouke a mere

The Boggy Creek bridge where it crosses U.S. Highway 71.
(Photo by the author)

roadside attraction in the scheme of the landscape. Given this, it's not hard to imagine what Fouke was like back when the monster reports first started coming to light. Not much has really changed, and that seems to be just the way the town prefers it.

SABBATH OUTPOST

The town may now be famous for a monster, but Fouke was originally founded by Reverend John F. Shaw as a religious colony in 1891.[6] Shaw had come to nearby Texarkana in 1876 where he became pastor of the newly established First Baptist Church and editor of the *Daily Visitor*, one of the city's first newspapers. At the time, Texarkana was a rapidly growing city whose businesses operated seven days a week to keep up with the demands of the local lumber industry and the non-stop influx of visitors and workers brought in by the railroads. As such, the town attracted more than its share of shrewd entrepreneurs, crooks, outlaws, saloon owners, and ladies of the night. Shaw found all of this to be at odds with good Christian values, most notably since the city itself and many of its people failed to observe Sunday as the Lord's Day. The gambling, hooking, and drinking ran every day in flagrant disregard for the Sabbath; even people claiming to be good Christians would often work on Sundays. Shaw sought to change the situation as best he could.

But even with Shaw's prominent position in the First Baptist Church, he was unable to influence a change in city-wide policy regarding observance of the Lord's Day. By the time he was up for re-election as pastor in 1879, several matters of dissatisfaction within the church and without led him to resign. So he founded the short-lived College Hill Baptist Church and later converted to a Seventh Day Baptist. After five more years of preaching against work on Sunday, Shaw ultimately concluded that perhaps Texarkana was not the best place for him and his flock. It was then that he made a decisive move to establish a religious colony outside the sinful grip of the big city.

To accomplish this goal, Shaw and fellow churchgoer John E. Snell purchased land from George Fouke 16 miles south of Texarkana. There they proceeded to layout the town just as the American frontier was coming to a close in 1890. Using his own self-published newspaper called *The Sabbath Outpost*, Shaw promoted the fledgling colony as a place where family, God, and education came first. In many regards, Fouke was a unique town that exemplified the pioneer spirit of early America along with the age-old struggle for religious freedom. It is worth noting that Fouke is one of the few towns in the United States that was started with the purpose of providing a religious haven. However, it would not be without its share of outlaw tales and controversy. It was, after all, a part of the Deep South during the formative years of expansion, a time that inevitably brought forth all kinds of unswarthy people — and apparently, hairy monsters.

BOGGYTOWN

People were said to have begun settling in the areas around what would later become Fouke as early as 1830. The creeks and rivers provided ample resources to establish small homes, and the steamships that began navigating the Red River in 1831 served to create an ever-growing lifeline of transportation. Wagon roads began to spring up as well, connecting small hamlet-type villages with the greater Texarkana network.

The coming of the railroad 50 years later brought with it a host of men eagerly seeking fortune. Among them was George Fouke, who arrived in 1875. He and his father-in-law had their eye on the abundance of lumber available in the Miller County area and sought to capitalize on it.

Mr. Fouke and his family took up residence in Texarkana, and within a decade he entered into business with two other men to form the enormous Gate City Lumber Company. The company, with its main plant based in Texarkana, manufactured all types of finished wood products for the growing construction industry. At the

Reverend John F. Shaw, founder of Fouke, Arkansas.
(Courtesy of the Miller County Historical Society)

time so many people were building houses and other structures that the lumber business was extremely lucrative. But in order to supply the raw wood needed to pump out so much finished product, the company had to rely on the resources from the outlying woods... including those near a little waterway called Boggy Creek.

In no time, the Texarkana, Shreveport, and Natchez Railroad line was extended into the woods, connecting several lumber giants, including Gate City Lumber, to the seemly endless natural resource. The railroad plunged into the dense woods, eventually terminating at Boggy Creek. According to the Miller County Historical Society, "Its waters would serve invaluably in the preservation of pine logs until they could be sawed up. The stream would also provide steam resources for the big sawmill and planer engines."

This "big sawmill" is the Boggy Mill, which was constructed on Boggy Creek for the purpose of manufacturing the lumber on the spot, as it was cheaper to transport finished pieces back to Texarkana than the raw logs. So from 1890 to 1904, the Boggy Mill served as the main source of materials for Gate City Lumber, making Boggy Creek's contribution to the industry a significant one.

Along with the construction of such mills, settlements known as "sawmill towns" would spring up. Like mining towns, these lumber towns would appear quickly and often disappear just as quickly when the gold—or in this case, lumber—dried up. But in the case of the Boggy Mill, which ran strong for approximately 14 years, a fairly solid establishment was able to flourish. Known as Boggytown, it not only had a catchy name, but it was large enough to warrant its own post office. According to the Miller County Historical Society, its population "varied between two hundred and five hundred during its fourteen year history."

The Historical Society also lists several prominent town members and mill workers, including planer manager Will Crabtree. The Crabtree family will later come to play a significant role in the Fouke Monster sightings, so it's interesting to note that they were living in Boggytown during that era.

Boggytown was located near the point where Boggy Creek crosses old Highway 71. Standing there today, it's hard to imagine that at one time a little town bustling with gambling establishments, saloons, hotels, stores, and a post office thrived at the lonely spot, since not a single trace of it remains. Apparently, the Arkansas landscape hides evidence of its ghost towns well and, presumably, its monsters, too.

And indeed I could find no historical sightings, or even rumors, of a hairy monster stalking Boggy Creek during the heyday of Boggytown, despite the increased human presence in the area. It wasn't until after the demise of the Boggy Mill in 1904 that people began to hear whispers that something strange may be lurking there. Perhaps it was coincidence, or perhaps it was the result of encroaching deforestation, that began to draw the monster out of the bottoms.

As I mentioned previously, Willie Smith went on the record claiming that his sister had seen the monster around 1908 in the vicinity of the old Boggytown location. After some further research and interviews with locals, I discovered that the sister Smith referred to is Kate Savell. I had heard of Kate's sighting, and when I mentioned Smith's claim, I was told that his sister and this woman were one and the same. I have spoken to other anonymous sources who claim that someone else—possibly a member of their own family—was the source of this sighting, dating it at around the same time, between the years of 1904 and 1910. Understandably, it's hard to pin down an exact date or a precise recollection a century later, but suffice to say, one or possibly more incidents occurred at the end of the Boggytown era that could be attributed to our mystery animal.

Another sighting occurred in 1916, although in this case it took place approximately 19 miles west of Fouke near Wright-Patman Lake in Texas. While it may seem like a considerable distance, it is worth noting that the incident took place near the origin of the Sulphur River, which is part of the larger waterway network that includes Boggy Creek. The story was reported by a retired geologist whose grandparents lived in a place called Knight's Bluff west of Queen City, Texas, and just south of the Sulphur River. Today there is still a campground on the edge of Wright-Patman Lake called Knight's Bluff, but the true Knight's Bluff was covered by water when the lake was created by the U.S. Army Core of Engineers in 1953.

According to the geologist, the incident happened on a summer night in 1916 when his grandmother was 18. She often told the story of what went on that night and the days to follow, so he and his family knew it well. On the night in question, his grandparents were returning home from town, navigating their mule-drawn wagon across the rough country roads. The moon was high and bright so visibility was good. As they turned onto the lane leading to their farmhouse, the mules began to act up as if they were spooked by something. Thinking that perhaps a snake was lying in the road, her father peered ahead, but saw nothing. A few seconds later they heard a strange noise coming from the east pasture, something like an eerie high-pitched wail or howl. The mules heaved as the family struggled to see what had made the noise. After a few moments, they saw a tall figure emerge from the dark line of trees adjacent to the field and walk out into the moonlight. His grandmother described it as being "tall or taller than a man and covered with long, dark hair." She also noted that "it stood absolutely erect and walked slowly toward them like man… not slouching like an ape."

As the creature moved across the field toward the wagon, it continued to howl, all the while motioning angrily at them with its long arms. By now the family was in a state of panic and shock. Seeing no other choice, the father reached for his rifle and fired once in the animal's direction. The shot presumably missed the creature, but the crack of the rifle was enough to send it running back into the woods.

The father then enlisted enough cooperation from the mules to get the wagon back to the house, where the family promptly jumped out and scrambled for the safety of their home. Once inside, they barricaded the doors and spent the rest of the night in a state of restless sleep, wondering just what it was that they had seen. The following day and a few days thereafter, the men of the family, along with a few neighbors, scoured the woods near the pasture searching for any signs of blood, fur, or tracks, but nothing was ever found. The family never saw the mysterious creature again.

Back in the vicinity of Fouke, a handful of smaller communities such as Jonesville, Fairland, Fort Lynn, Lizarlope, Corinth, and others managed to outlive Boggytown and maintain a small-but-steady population of residents in the countryside. Accounts of the mystery monster during this time are hard to come by, but I did learn of one more which occurred around 1932. The story was told to me by John Attaway, a current Fouke resident. It involved a friend of his by the name of Ace Coker, who was living with his sister in a house near Fort Lynn, three miles due south of the old Boggytown site. As the story goes, he was sitting on the porch one day, while his sister was hanging fresh laundry on the clothesline.

At some point he got up to go back inside the house. As he opened the screen door, he tilted up his hat, and there standing near the porch was a large, hairy man-like animal. It had apparently been able to slip up to the porch without making a sound and was now eyeing him curiously. Startled, Coker hurried into the house before looking back again. When he turned around, the thing had already moved out by the fence on their property, and from there it slipped out of sight.

There may have been other encounters in the first half of the 1900s, but since many people kept these types of things quiet and there was no media coverage, many tales have probably been lost. Attaway said that Coker did not like sharing his story, so presumably even this was not known about beyond a small circle of friends.

It would be several decades before stories of a monster began to register on any significant scale with the locals. But as the destruction of wildlife habitat continued in the wake of the lumber

A group of happy citizens in front of the Fouke post office circa 1921. (Courtesy of the Miller County Historical Society)

boom, it was only a matter of time before the creature emerged to haunt the growing number of intruders.

Workers during early construction of Highway 71 in 1928.
(Courtesy of the Miller County Historical Society)

3. Jonesville Monster

The Haunting Begins

The Fouke Monster did not start out as the Fouke Monster. Well, that is to say it *did* start out as a "monster" and still is one (as far as we know), but in the process of its rise from the swamps to greater fame, a renaming took place. It actually began its cryptid life as the "Jonesville Monster," named in reference to a small community located about six miles southwest of Fouke.

If anyone thinks the area around Fouke where Boggy Creek crosses the highway is spooky at night, then just take a drive down the backroads of Jonesville late one evening. This is a true monster's paradise where the trees bend threateningly toward the road, and the brush is so thick it can hide even the largest of predators that may be lurking only a few feet away.

Jonesville was founded in the late nineteenth century by a family of settlers bearing the Jones surname. Many of their descendents still live there today, some of which have given me insights into the monster's tale. The community itself is not much larger than it was back in the early 1900s, and it still enjoys a low profile setting amid the rich Sulphur River Bottoms. The borders of the community fit roughly between Boggy Creek on the north, Highway 71 to the east, and the Sulphur River to the southwest. Just a few miles to the west sits the magnificent Sulphur River State Wildlife Management Area while Mercer Bayou spreads out just beyond the Sulphur River to the south. All in all, Jonesville has a bay window view of some of the state's most remote and inhospitable swamplands. When the rivers flood, all sorts of game are forced to expand their territory to avoid the mucky wallows. It's times like these when the residents of Jonesville watch closely for any strange figures lurking in their backwoods.

Taking a short jog south on the old U.S. 71, Jonesville arrives without fanfare. In fact, it is easy to pass right on by if you don't know where you are going. There's only one road in and one road out of Jonesville. This main thoroughfare is marked by a stark white church sign, and other than a few dirt driveways that disappear behind the trees, there's not much else to announce its presence. Turning down the road and heading into Jonesville, all semblance of highway is immediately lost. The modern world seems to disappear in the rearview mirror like some strange Twilight Zone episode. The roads here are crudely paved and without curbs, as you would expect from old country roads. The small ditch that slopes off each side of the road is quickly met with wire fencing, trees, and brush, lining both sides like a natural wall. You can't see the houses that dot the countryside. You know they are there because of tell-tale driveways winding behind the trees, but it's hard to make out any real details even in daylight. The foliage is thick, especially when summer is in full bloom. And if it's hard to see a house, imagine how hard it is to see a dark hair-covered creature that may prefer to watch *you* instead of letting its presence be known. Driving in the area always makes me realize how easy it would be to hide in the dense underbrush just out of sight.

As the evening sun continues to slip down behind the treeline, a spooky calm starts to set in. The sun's errant rays still break through the dense forestry here and there, as shadows begin to form pockets of dusky darkness all around. In many places the trees are so thick they envelop the road in a complete canopy. Rounding a corner and proceeding down an incline into these long black tunnels is like driving into the great maw of an enormous beast. It's definitely a creepy place, where the imaginations of passersby and tourists could certainly run wild. Maybe *they* could mistake the shadowy figure of a distant hunter for that of a hairy monster, or believe a bear to be a bipedal manimal. But it was *not* tourists or city folks who originally encountered the legendary beast of Boggy Creek; it was the locals, the people who have lived here all their life. These people do not paint the dark woods with fancy metaphors or pay heed to gaping maws of canopy trees. These country folks know

The original settlers of Jonesville, circa 1895.
Seated L to R: Mrs. Bose Jones (holding Isadore), Mr. Bose Jones.
Standing: Albert "Bank", Mary Jones Grandberry Davis, Ed,
Epsie Jones Harris, Commadore.
(Courtesy of the Miller County Historical Society)

well the wildlife of the area and regard this place as their home, not a spooky place of legend. They have little time for nonsense as they carve out an existence in the often inhospitable setting. So when the people of Jonesville first began to speak of a strange ape-like animal that was haunting their land, it might have done us well to listen.

Sifting through old *Texarkana Gazette* newspapers, I came across what appears to be the earliest reference to a sighting in modern times. It came from Leslie Greer who served as Miller County Sheriff from 1967 to 1974. In 1946, Greer states in the article, "I was campaigning for tax assessor and stopped to talk to a lady sitting on her front porch. She lived about halfway between

Fouke and the Below Bridge.[7] She told me that she saw some kind of animal go down in the field in a low, bushy place. She said it looked kind of like a man, and walked like a man, but she didn't think it was a man."

Greer never gave much thought to the report until the rash of 1971 sightings, so up until that time there was little awareness of the creature beyond the boundaries of Jonesville. But as with Willie Smith and Sheriff Greer, once the frenzy started at the Ford house, all the strange incidents from the past began to make sense, at least in the eyes of those who already possessed a piece of the scattered puzzle.

Another person alleged to have early knowledge of the Jonesville creature was James Crabtree. In about 1955, he claimed to have come up on a bizarre creature while fishing in a nearby river. The creature was large and hairy like a gorilla, but it walked like a man. I confirmed the story with Crabtree's great niece, Syble Attaway, who kindly agreed to meet me one afternoon in Fouke. She was approximately 10 or 11 years old when she heard her great uncle tell about what he had seen that day.

"We went over there one day," Ms. Attaway began, describing how she had gone to her grandparent's house to visit a sick relative. "They would be sitting around talking, you know. Uncle James said, 'I went down to the river that morning to run my trot lines. I was floatin' down the river and saw something up on the bank.' He said 'it was sittin' down like a man, washing its feet. I tried to ease up, or float up on it, but it stood up and walked off.'"

As the adults talked, Syble's grandfather, Lee Crabtree, suggested that it might have been a gorilla. But James was sure it wasn't. The size of the animal didn't seem right. Lee thought perhaps his brother had seen a bear, but again James disagreed. She recalled his unsettling answer: "It walked like a man; it didn't walk like a bear. It just stood up and walked off."

It seems unlikely that an experienced woodsman like James Crabtree would mistake any sort of local wildlife for a mysterious creature. The locals who knew James, or knew of him, believed he was one of the best hunters and trappers to ever come out of

One of the roads near Jonesville, pictured in 2011.
(Photo by the author)

Jonesville. It just came down to whether one believed Mr. Crabtree or not. Ms. Attaway herself felt that the story was true. "There was something [to the stories], I guarantee there was. They just didn't sit around and make things up. They might exaggerate a story about an old coon dog or something, but they didn't just sit around and make up stuff that didn't really happen. These were just three old people sitting there discussing their day to day life."

Though she heard this account many years ago, Ms. Attaway remembers the tale distinctly. Not only must it have sounded strange and frightening to a small girl, but her grandparents told her not to repeat the story for fear that outsiders might take them to be crazy. In fact, this was a real concern and one of the reasons that these early Jonesville sightings did not make the newspaper. No Jonesville resident in their right mind would call up the *Texarkana Gazette* to report a "monster sighting." Back in those days, these country folk did not have much in the way of riches, all they had of value was their *word*. This being so, the integrity of their word was highly important. To sully it with tall tales of monsters was not something most people were willing to risk. Not only did they run the risk of losing the value of their word, but a one-way ticket to a sanitarium might be in their future if they continued to purport such fantasies.

Despite such hazards, however, the strange stories did make the rounds, at least locally. According to the recollections of Fouke residents, such as Frank McFerrin who was born and raised there, James Crabtree's story circulated among the locals when it first occurred in the 1950s, but was not taken seriously at the time. "This was one of the first sightings that received a wider acclaim. In other words, it reached up here to Fouke. But I think many people wouldn't put credibility in it at first," McFerrin told me as I interviewed him one afternoon in Fouke. I asked if the newspapers had gotten wind of the incident. McFerrin explained that: "[News of] the sighting didn't travel very far. It was just something local people talked about."

But James Crabtree wasn't the only one who had seen a strange animal lurking around Jonesville at the time. Ms. Attaway recalled another incident that took place not long after her great uncle's sighting on the river. She was riding in the car with her grandparents, heading back to their house, which was located near Independence Cemetery on County Road 9. "We were going home at night and I was lying down in the back seat. It was probably after seven o'clock because it was bedtime," she remembered. "We had almost gotten to grandma and grandpa's when something crossed the road in front of them." Syble didn't get a look at the creature, but she will never forget the strange conversation that followed.

"Lee, what was that?" her grandma asked, concerned as to what it might have been.

"I don't know, but it sure was hairy," he responded.

"It walked like a man."

Her grandfather paused a moment and then said: "It had too much hair to be a man."

A Change of Fate

One of the most significant incidents involving the alleged creature occurred in 1965 when an encounter with Lynn Crabtree, a boy of 14, would forever link it to the legendary Crabtree family and ultimately change its destiny.[8] Without this chance meeting, Charles Pierce's *The Legend of Boggy Creek* may not have turned out as well as it did on the big screen, since the Crabtree's stories would go on to play a major role in the movie, due in part to the notoriety of this incident.

Though many books on the subject of Bigfoot have summarized the incident over the years, the best and most thorough account can be found in the first memoir by J.E. "Smokey" Crabtree, Lynn's father. Simply titled *Smokey and the Fouke Monster*, the book goes into great detail about this monster sighting and all the events that followed, including the monster hunts and the filming of *The Legend of Boggy Creek*. The book, which includes the fascinating tale of Smokey's upbringing and adult life, not only makes for good reading but is inspirational as well. Though Smokey Crabtree's stories were often damaging to his own family's well-being and privacy, they are woven into the monster's very existence. As such, one cannot tell the story of the Fouke Monster without including the tales of Smokey Crabtree.

Here is how it all began…

It was early in the evening as the sun began to set over Jonesville. Lynn Crabtree, a seasoned woodsman and hunter even at his young age, set out to hunt some squirrels near the lake on the Crabtree property.[9] He was armed with a 20-gauge shotgun loaded with squirrel shot, which would be sufficient for small animals but not for what he was about to witness.

As he sat by a tree waiting for any unlucky squirrels to happen by, he heard the hoofbeats of horses making their way down an old logging trail and eventually splashing into the waters of the lake. The horses belonged to a neighbor and often ran loose, so it was nothing out of the ordinary. He then heard the bellowing of a dog from the same direction. With this, Lynn got up and began to head toward the sound, thinking that perhaps their dog might have been injured or gotten hung up in a fence. But as he approached, he realized that the noise must have been coming from the *thing* that now stood in front of him, a mere 30 steps away. It was some type of "hairy man or gorilla type beast with very long arms." Lynn could only assume that it had been the reason the horses fled into the safety of the lake.

The beast, which seemed agitated, suddenly stopped as it caught sight of Lynn. The boy could see that the creature stood an estimated seven or eight feet high with reddish-brown hair about four inches long. Its face was obscured by hair with "only a dark brown nose showing, flat and close to his face." Thinking it must surely be a man, Lynn raised his gun in an attempt to frighten him off. But the strange manimal didn't react to the danger posed by the gun and started walking toward the boy. Surprised and scared, Lynn shouted a warning before firing off a round. He aimed for the head, but the beast seemed totally unaffected as it continued to advance. The Crabtree boy shot off two more rounds before finally fleeing in panic towards his parents' house.

If the Crabtrees had been the only ones present when Lynn returned from the woods, then the story might have ended there. But as circumstance would have it, the family had company at the time, so when the horrified teenager burst into the house, the story was fated to become public knowledge.

After Lynn had calmed down, Smokey made his way out to the very spot where the boy said the encounter occurred. He found evidence of the gunshots, which had hit the tree trunks at a height consistent with firing at the head and torso of a seven-or-eight-foot bipedal creature. Surprisingly, however, although the family had heard three shots, just as Lynn described, Smokey could not find any spent shotgun shells.

Puzzled, Smokey returned to the house. The story was undeniably fantastic, but there was no reason for anyone at the house to question Lynn's sincerity. In Smokey's own words, "The community knew [Lynn] well... and respected him highly for being a truthful boy." As a boy who was raised in a very traditional and moral fashion, he was never known to be a liar or to make up ridiculous tales. The Crabtree's visitors knew this as well, as the family was known to be honest and hard working.

I never did get a chance to interview Lynn about the incident, but I spoke to several people who had known him over the years, including his lifelong friend Mackey Harvin who also grew up in Jonesville. He confirmed the details of the event and spoke highly of Lynn's credibility. It was his opinion that if the young man said he encountered such a creature, then he absolutely did.

There is, of course, no evidence to back up Lynn's tale. So what are we to make of his story? If he wasn't lying, and if it wasn't a case of misidentification, then one can only conclude that there was some kind of mysterious ape-like creature living in the woods near Jonesville.

In light of this, I've often pondered the two most puzzling aspects in the story: first, that there were no spent shells on the ground when Smokey returned to investigate, and second, that a boy who could shoot a squirrel with his firearm could not hit a target that would have been at least two to three feet wide across the shoulders. If the boy's estimate that he was less than 30 steps away is correct, and given that a shotgun requires less accuracy than say a rifle or handgun, it would stand to reason that he could have easily hit the animal at least once. Yet, the thing kept advancing on the boy and no blood was found at the scene.

If the incident did in fact occur, one must assume that the creature either picked up the shell casings or scattered them out of sight before leaving the scene. Perhaps the creature had been curious seeing those shiny objects on the ground.

As for Lynn missing the creature with three shots, I tried to think back to when I was a boy of 14 hunting the backwoods of Texas. Granted, I was probably not as seasoned as Lynn, who had spent his whole life in a rural setting, hunting animals to put food on the dinner table. But regardless, I can imagine the cold steel of my grandfather's 20-gauge Remington shotgun in my hand shaking

1965: Lynn Crabtree encounters the creature at Crabtree Lake.

as I raise it in warning, while a hairy, seven-foot "monster" begins to move in my direction. I had missed shots at my dad's Coors cans while practicing under calm circumstances, so I can see how the boy might have missed, his hands shaking like jackhammers as he pointed it toward the hairy thing, not sure if it was man or ape. Or maybe Lynn did score a hit or two, but the small size squirrel shot had not penetrated the creature's thick fur, which is certainly possible.

Beyond such conjecture, the mystery remains. Lynn *never* spoke to the press about the incident. He passed away in 2011.

OTHERS COME FORTH

Once the word about Lynn's frightening encounter got out, it wasn't long before other creepy tales came to light to back up the boy's story. First, Lynn's great uncle, James Crabtree, reminded everyone that he had also seen something that fit a similar description in the nearby woods 10 years earlier. Fred Crabtree also came forth with a similar story saying that he had seen a hairy man-like creature prowling in the dense backwoods near Jonesville months earlier. He was out hunting when he came across the shadowy creature as it stood among the thick trees. Like James, he was unable to get a good look before the thing slipped out of sight, but it seems reasonable to assume that it was the same creature or perhaps another of its kind. He hesitated to shoot, since it was impossible to determine if it was man or animal. Circumstantial and nepotistic as these tales were, they still gave credence to the boy's extraordinary claim that a monster was living among them.

An especially chilling account came from Mary Beth Searcy who also lived in the area of Jonesville near Boggy Creek, about three miles from the Crabtree property. Mary Beth, a teenager at the time, was spending the night at home with her mother, older sister, and baby niece, while her brother and father were spending the night elsewhere. When the other women went to bed early with the baby, Mary Beth continued doing her schoolwork. As the spring air cooled with the deepening night, Mary Beth's sister asked her to cover the bedroom window to prevent the baby from becoming ill. Grabbing a blanket, Mary Beth approached the window and proceeded to cover it. As she did so, she glanced into the yard, which was bathed in enough sparse moonlight to see the immediate area between the house and the outlying trees. She was shocked to see a large, hairy creature emerging from the woods, walking on two legs as though it were human. She screamed and ran from the window as the other women bolted awake. The women spent the rest of the night in sleepless terror.

This account would later be recreated in the movie *The Legend of Boggy Creek* with Judy Baltom playing the part of Mary Beth, and Mary Beth playing the part of her own sister.[10] According to Fouke residents, the movie's portrayal of the events is an accurate retelling with the exception of the dead cat shown at the end. (It apparently died of fright.) If true, this must have been a horrifying experience

The Searcy house—one of two structures used in the movie that still stands today.
(Photo by the author)

for the young girl, since they had no phone and the nearest neighbor was more than a mile away. The Searcy house still stands today, although it's been vacant for many years. And even though neighbors now live in the homes next door, it's still something of a spooky sight; it's not hard to imagine what it must have been like back then when Mary Beth saw the creature leering in the darkness.

I tried many times to interview Mary Beth, but she was not willing to talk. Over the years she's experienced too much ridicule about the incident and had just decided to stop talking about it. The last time she commented publicly was in 1992, when she was interviewed by a reporter from KTBS-TV in Shreveport, who was doing a twentieth year anniversary retrospective on the movie. On camera, Mary Beth explained how she felt:

> Mary Beth: "They ask you about it, you know, like they're really serious and then they laugh at you [saying] 'ah, you don't believe that do you... you don't expect us to believe it??' Well, if hadn't expected them to believe it, I wouldn't tell 'em. I've never told a lie in my life."
> Reporter: "But you know it's real as far as you're concerned."
> Mary Beth: "I saw it."

Her words and the sincere tone of her voice leave little doubt that she still sticks by her story.

After the news spread of Lynn Crabtree's sighting, men started showing up at Smokey Crabtree's house offering to help hunt down the creature. Nearly 20 locals, armed with guns, horses, and tracking dogs scouted the area, but they were unable to find any solid evidence of the creature's presence. One of the men on horseback claimed that some kind of creature had passed by and spooked his horse, but he was unable to get a good look at it.

Smokey continued to look for the creature on his own for the next few nights, but again, the experienced woodsman came up empty-handed and somewhat disappointed that he was unable to conclusively prove Lynn's story. But there was a glimmer of hope that a real living creature was out there. In a 1971 interview with the *Texarkana Gazette*, Smokey told how they were able to solicit a response from the monster:

> "At that time we tried to get dogs to run the thing but they wouldn't go in the woods. For three nights after that we used a wounded rabbit call and got the creature to answer it but he would never get close to us.
>
> "It started making a noise like a house cat then went into a chatter something like a goat then into a noise something like an owl. After a while it seemed to get annoyed and made a sound like a cross between a scream and a growl. That sound would really make your hair stand on end."

A short time later, another 14-year-old boy from the Jonesville area, Kenneth Dyas, claimed to have encountered the monster while deer hunting. Dyas is now deceased, so very few details can be found regarding this incident. According to sources, Dyas shot at the creature and fled the woods as fast as he could.

Another sighting report came to me via Frank McFerrin, a long-time Fouke resident and curator of the Miller County Historical Society Museum. He estimates that the incident took place around the same time, in 1964 or 1965, although the details of Phyllis Brown's spooky encounter with a similar creature were not known publicly until several years after the fact.

As the story goes, Ms. Brown was deer hunting southwest of Fouke within an eighth of a mile from the Sulphur River. This would have placed her in the general vicinity of the other Jonesville sightings. McFerrin recalled: "She was in a deer stand and she heard the dogs running what she thought was a deer. She looked up a pipeline and there came what she thought was a black guy... running upright. He was far off at first, and then she got to realizing 'no that's not a black guy, that's something that's got hair on it that's black... running like a man!' The thing was moving in a northwesterly direction behind her, and when it passed by she got a good look at it. She said 'it was not something she had ever seen before.'"

At the time of the sighting, she only told a few family members and trusted friends what had happened as she was reluctant to tell her story for fear of ridicule or damage to her reputation in the community. Even when *The Legend of Boggy Creek* was being shot in 1971-72, she declined to tell her story to director Charles Pierce. When I asked McFerrin about her credibility, he replied, thoughtfully: "I do know people that are credible that have seen something they could not explain. I don't think she made this up. I think she literally saw something."

A few years later, around 1967, Charlie Walraven was driving toward his home on County Road 9 late one evening. As his old '52 Ford coughed its way down the lonely stretch of rural road, a man-like creature suddenly darted out in front of him. It was only visible a few seconds before it disappeared into the thick grove of trees on the opposite side of the road, but he had seen it long enough to tell it was something running on two legs. The beast was hairy and fit the profile of the Jonesville Monster.

A year later, sometime in 1968 or 1969, a similar beast was seen on two separate occasions just two miles south of Jonesville as it searched for food around the home of Louise Harvin. Ms. Harvin is now deceased, but I was able to speak with her son, Mackey Harvin, who recounted the events. The first encounter occurred at sunrise one morning as Ms. Harvin prepared for work at the D&W Packing Company where she was a meat cutter. Hearing a ruckus in her hog pen, she went outside to check on her animals. As she headed across the yard toward the pen, she was startled to see a large, hair-covered animal standing on two legs inside the fence eating from a pile of scraps that had been left out the night before. She said the animal had long reddish hair that covered its body while its face was a darker shade of brown with a large, flat nose. Food scraps hung from its mouth and clung to the fur of its chest and hands. When it saw the woman, the creature immediately stopped eating. Within seconds, it jumped the fence and disappeared into the woods. Startled, but not overly frightened by the strange beast, Ms. Harvin continued on to work, although at the time she told no one what she had seen.

The creature made a second visit not long after the first incident. On this occasion, Ms. Harvin was stepping out onto the porch early one morning when she surprised the animal as it was licking remnants of food from an old hub cap that was being used as a dog bowl. As before, the creature paused and looked at her. Ms. Harvin made note of the creature's scraggly hair and its ability to stand on two legs. She could also see that it had rather long canine teeth, which resembled those of "a baboon, although the face was much more flat like that of a gorilla." After a few seconds, the animal bounded off into the woods, never to be seen by Ms. Harvin again.

Her son Mackey, who was off serving in the war at the time, did not hear of his mother's encounters until a few years later when he returned. He was already familiar with the sighting by his friend Lynn Crabtree and the incident at the Searcy house, so he was not completely surprised that his mother had also seen the creature. Given the creature's flat, brown nose and overall hairy body, he feels quite certain that his mother had been surprised by the same creature, or one of its kind, that Lynn had seen. Mackey did not feel his mother would have any reason to make up such a wild tale. But like so many others, she was reluctant to tell most people that she had seen the Jonesville/Fouke Monster. The Harvins had seen first hand what had happened to Lynn when his story got out. He had suffered considerable ridicule at the hands of his classmates and other locals. Ms. Harvin was not eager to experience the same.

Yet another sighting of the mystery beast occurred in 1969, although the details of this case are somewhat thin. According to Sheriff H.L. Phillips, two coonhunters were hunting near Jonesville when they proceeded down into a draw. As they reached the bottom of the short incline, they turned and noticed something standing behind them. It was a large upright creature on two legs and covered with bushy hair. Just how the men reacted is not known, but presumably they either retreated up the draw or the creature slipped away into the woods before they could get a better look. The creature was often surprised by encroaching humanity, but it was easy enough to slip back into the shadowy woods of Jonesville. For the creature... it was a perfect haunt.

1968-1969: The creature prowls around the Harvin home, two miles south of Jonesville.

THE STRANGE RUNNER

One of the most astounding reports that seems to corroborate the stories told by the Jonesville residents didn't surface until long after Pierce's movie had been made. One of the witnesses in this case was Carl Finch, guitarist and founder of the popular polka/rock band, Brave Combo, from Denton, Texas. Founded in 1979, the band has gone on to enjoy a wide following and legendary status with a slew of honors including two Grammy awards, an appearance on *The Simpsons* cartoon, and a cameo in the 1995 feature film, *Late Bloomers*, among other achievements. I had the good fortune of speaking with Finch, who still lives in the Denton area.

According to Finch, the incident occurred late one night on a lonely stretch of Highway 71. It was the spring of 1967. Earlier that evening he and his group at the time had performed at a "Battle-of-the-Bands" in Shreveport, Louisiana. After the show, he and his cousin made the late night drive from Shreveport back to Texarkana—where he was living at the time—in her Volkswagen Beetle. This trip took them right through the Jonesville/Fouke area. As they traveled along 71, which would have been extremely desolate and dark back in the late 1960s, they noted an upright figure in the headlights as the thing moved alongside the road in front of them. It was traveling in the same direction as their car, and as they got closer they could see that it was running at a fairly rapid clip. Finch's first impression was that it must be "a guy in a brown coat." That seemed rather odd, however, as this was a very dangerous and unlikely place for a man to be out jogging late at night. Besides, the weather at the time did not warrant a heavy coat, especially if one intended to work up a sweat.

As they closed the gap, it became increasingly apparent that there was something more bizarre about the late-night runner. "We noticed that it had really long arms and fur... not a coat," Finch stressed. "It was well-defined and running very fast with a gait that didn't seem human." The creature did not react to the car's presence but kept on moving quickly down the road as they sped by in the car.

Once they had passed the figure, the headlights no longer provided illumination, so it was impossible to get a better look at the face. "We were too scared to stop," Finch remarked. "We just kept driving." Once they passed the creature—or whatever it was—it simply faded into blackness. And eventually into a distant memory.

It wasn't until the mid-1970s, when Finch caught a showing of *The Legend of Boggy Creek*, did he recall their sighting. Seeing the story of the Fouke Monster on the big screen and learning about its connection to the lonely stretch of Highway 71 gave him pause. Was it the Fouke Monster he had seen that night years ago? Finch believes it's a definite possibility. How else could you explain a jogger running down a dark, desolate highway in a fur coat well beyond the witching hour?

This event is significant to the Jonesville-era monster since Finch and his cousin were third-party witnesses. They did not live in Jonesville or Fouke, and at the time had no knowledge of what had been going on there. "I didn't know about Bigfoot or the Fouke Monster at the time," he told me. So if the residents of Jonesville had all been caught up in some mass hallucination or hoax, Finch was certainly not a part. Yet the creature he describes seeing that night matches very closely with what locals have described, making it hard to dismiss the whole affair out of hand.

It is a widespread belief that the monster had only been seen a few times prior to the Ford incident, but as these reports show, there is a long history of sightings of the creature before 1971. It's just that the folks down in Jonesville didn't see this as big news, so they only shared their experiences with the people they knew and trusted. The presumption that a strange creature was living in their midst was just another part of daily life. They had, on occasion, organized hunts in an attempt to solve the mystery, but beyond that there was not much they could do. After all, it had never actually hurt anyone, so if it wanted to share their lonely little bayou, then so be it.

Brave Combo founder and frontman, Carl Finch.
(Photo by Jane Finch)

4. From Swamp To Big Screen

Enter: Charles B. Pierce

Whenever someone brings up the classic 1972 horrordrama *The Legend of Boggy Creek*, it never fails to produce at least one response along the lines of "that scared the hell out of me when I was a kid!" And deservedly so. For anyone old enough to have seen the film during its heyday at a vintage drive-in, or later on television or video, it spawned a host of reactions, including nightmares, fear of the woods, fascination with ape-like creatures, anxiety when sitting on toilets in old houses at night, and/or a foolhardy desire to chase "real monsters" in the swamp... just to name a few. I was fortunate to catch *The Legend of Boggy Creek* at an old Texas drive-in as a child, so I can attest to the validity of these reactions. For me and countless others at the time, the tale of Boggy Creek was more than just a legend... it was real! And it was the movie, ironically, that helped establish in our minds a more permanent reality for the Fouke Monster.

The Legend of Boggy Creek was the directorial debut of Charles B. Pierce,[11] who ultimately put aside a career in advertising to pursue filmmaking. Though *The Legend of Boggy Creek* would end up being one of his most memorable achievements due to its monetary success, classic scares, and influence on future filmmakers, he did direct other films such as *Bootleggers*, *Winterhawk*, *The Winds of Autumn*, *The Town That Dreaded Sundown*, *Grayeagle*, *The Norseman*, *The Evictors*, *Sacred Ground*, *Hawken's Breed*, *Chasing the Wind*, and eventually the less-than-classic sequel, *Boggy Creek II*. In addition, Pierce's credits include being an actor, producer, and writer for his own films and others. One of his writing credits includes the screenplay for the 1983 Dirty Harry film, *Sudden*

Impact, starring Clint Eastwood. Remember the line "Go ahead, make my day," which was imitated by virtually every jokester back in the 1980s? Well, that was conceived by Pierce!

In fact, as I have learned through my research, Charles Pierce was a very talented guy, not just in films, but in graphic design, illustration, news casting, writing, acting, singing... you name it. He was a virtual renaissance man who defined the DIY (do-it-yourself) attitude of an industrious artist. This kind of thing might be common now with the hordes of indie filmmakers shooting in their backyards direct to DVD, but back in the day, Pierce was a pioneer.

Pierce passed away in March of 2010, but I had the pleasure of speaking to his daughter, Amanda Pierce Squitiero, about the man and his legend. She was no stranger to his movie sets and even appears as a small girl in the climax of *The Legend of Boggy Creek*. It was surreal telling her how her father's movie had influenced my own family back in the 1970s. After seeing the movie, I can remember countless times when my dad would growl and jump out of the shadows to scare us. My mom and sister would scream "the Boggy Creek monster!" while I laughed.

But for Amanda it was much more emotional. She recalls being frightened nearly to death as she was shooting the scene in the movie. The fear she had on screen was real that night, back in 1971, in that old frame house out in Fouke. When she spoke to me about her father, her emotional ties to him were evident, along with the respect she has for his accomplishments, not just with *The Legend of Boggy Creek*, but throughout his life.

Charles Bryant Pierce grew up in Hampton, Arkansas. From a very early age he was involved with film. As a kid, he and his lifelong friend, Harry Thomason, who would go on to create popular television shows such as *Designing Women* and *Evening Shade*, used an old 8mm camera to make their movies. It must have been here that the cinematic seed was planted, although as a young man he set his sights on a career in graphic design. By his mid-twenties, he was working as the art director for KTAL-TV in Shreveport, Louisiana, but his do-it-all spirit would eventually land him in other positions, including weatherman and host of a children's cartoon program for the station.

In 1969, Pierce moved to Texarkana where he opened a small advertising agency, while at the same time, played a character called Mayor Chuckles on a local television show. Although he was popular in his role as Mayor Chuckles, it was the advertising work that gave him the opportunity to use all of his combined talents, and thus set the stage for his imminent jump to filmmaking. Using a basic 16mm handheld camera, he began producing commercials for local businesses, one of which was Ledwell & Son Enterprises, a builder of 18-wheel trailers and other farming equipment. Ledwell commissioned a series of commercials that gave Pierce the opportunity to film heavy trucks on the highways and other machinery in the fields. This was an important step, since it was Ledwell who would later put up the money to film *The Legend of Boggy Creek*.

Like all the residents of Texarkana in the early 1970s, Pierce had been following the sensational newspaper reports describing a hairy, ape-like creature which haunted the creeks near Fouke. As a result of his interest in the stories, he conceived an idea for doing a regional film based on the phenomenon. Seeing an opportunity to capitalize on the frenzy, he approached his wealthy client, Mr. Ledwell, and presented the idea. But Ledwell did not respond with much confidence in Pierce's ability to make a film. He was also doubtful of the subject matter. In a 1997 interview with the popular horror magazine, *Fangoria*, Pierce recounted his first proposal to Ledwell. When asked what kind of movie he wanted to make, Pierce told Ledwell: "I want to do one about that booger that's jumpin' on folks down the road there. It was in the papers every morning, you know, about the Fouke Monster and how it'd jumped on somebody else."

When asked at the time if he believed in the monster, Pierce simply replied, "I don't know if I believe it or not—but it sure will make a good movie!" Ledwell finally agreed and signed on to back Pierce.

With Ledwell on board, Pierce then needed to enlist the cooperation of the Fouke residents. This would prove to be even more difficult than convincing the stodgy old business man to bankroll a Bigfoot movie. When Pierce made his first scouting mission down to Fouke, he asked some of the locals what they thought of the idea. On the whole, the folks of the little town did not take kindly to the notion. Pierce remembers: "They didn't want to make a movie. They didn't ask for money. They didn't want anything. They wanted to be left alone, in fact."

That would later become a huge problem for Pierce, but it did not deter him one bit. The story was too good, and if the people of Fouke did not want to help propel their namesake monster to the big screen, then Pierce would have to work around that. And he did.

Pierce continued to interview locals until he found some who didn't mind sharing their stories or helping out with the film. One man offered to take Pierce down to his barn where he told of a startling encounter with the beast. The man had apparently surprised it one morning as he stepped outside to start his daily chores. He had a long look at the thing as it headed back out across a pasture toward the cover of the trees. Like other witnesses, he described it as a hairy, man-like creature that walked upright on two legs. Pierce was enthralled with the story and wanted to include the scene in his movie, but the witness wanted no part of the film, whether reenacting the encounter or simply mentioning his name. Pierce agreed to the anonymity, knowing that it wouldn't matter whether it was the actual storyteller or someone else who re-enacted the scene in the movie.

After more research and story-gathering, Pierce felt he had enough compelling source material to move to the next step, which was to find a competent writer to pen the screenplay. For the job, he hired an advertising associate by the name of Earl E. Smith, who also lived in Texarkana. Smith had never written a movie script, but like Pierce, he would prove to be just the right man for the job.

Using the working title "Tracking the Fouke Monster," Pierce went ahead with filming despite the fact that a script had yet to be drafted. Since they planned to use a documentary-style approach with narration overdubs, it would not require much in the way of rehearsed dialog, which was perfect since he had no real actors

*Pierce signals the crane operator to raise the camera platform while
filming a high angle shot.
(Courtesy of the Texarkana Gazette)*

on board to even deliver the lines. Instead, he planned to film
actual Fouke residents as they acted out the monster encounters,
which had appeared in the news or he had learned about through
personal interviews. One such local was Smokey Crabtree, whose
son (Lynn) had come face-to-face with the monster six years earlier.
While seeking information about that encounter, Pierce and Smith
also talked Smokey into acting as sort of a tour guide around town
and in the swamps. This was a fortunate move, as Smokey's lifelong
experience in the bottoms proved to be invaluable when it came to
accessing the remote regions for filming.

In exchange for a nominal fee, Smokey agreed to guide Pierce
through the waterways of Mercer Bayou so that he could shoot
the scenery. Each morning, Smokey would launch his handmade
canoe into the bayou, rowing Pierce around until the director got
the shots he needed. Pierce also shot footage of Travis, Smokey's

other son, who stood in for Lynn since he did not want to participate in the movie.

In some cases, Smokey also acted as an informal liaison between the movie makers and the townsfolk, convincing locals to participate in the movie, especially those who claimed to have seen the monster and had firsthand stories to tell. Pierce also resorted to some crafty low-budget ingenuity to cast the rest of the parts: he simply hung out at a local gas station and waited for people to drop by. When he spotted someone that fit the description of a person he needed, he would ask them if they wanted to be in a movie. Pierce described the process in the interview with *Fangoria*:

> When someone pulled in, we'd say, 'Now she'd be a good Peggy Sue.' And we'd walk out there to the gas pumps and say, 'Ma'am, we're shootin' a little movie. Would you like to be in it?'
>
> She'd say, 'Well, what do I have to do?' And I'd say, 'Oh, you just run across this field out here.' We'd get out there in the field and she'd say, 'What do I do?' and I'd say, 'I want you to come across over there screamin', and run real fast. We never did makeup or any of that, unless the creature got on 'em and then we'd put on a little ketchup for blood.

Eventually Pierce had enough Fouke residents on board to make a go at it. Some were even eager to participate in the film. The chance to potentially see themselves on the big screen compelled them to show up and try their hand at acting. Even the local police force assisted Pierce by setting up roadblocks and supplying props such as vehicles.

For a crew, Pierce relied on volunteers. He already had a small roster of high school age kids from Texarkana who had helped him shoot commercials, so he called on their help down in Fouke as well. Pierce rounded out his crew with some of the local Fouke kids and other volunteers. Since the kids often had school or chore obligations, they were not always available every day. Having a large pool of volunteers ensured that he would have enough help during the shoots, although the ever-changing makeup of the crew did not make things easy.

In the role of the monster, Pierce used a few different actors, including his brother-in-law, Steve Lyons; Steve Ledwell; and local boy, Keith Crabtree, who ultimately received the credit in the film. To create the costume, Pierce used a bit of ingenuity to fashion something that would reflect what people claimed to have seen and would also hopefully frighten audiences in the climatic final scene.

> Once we got to the ending, I knew we had to do something for some kind of payoff. So we ordered a gorilla suit from some costume house in Los Angeles, and I went down to the five-and-dime store and bought a bunch of old wigs, and we cut 'em into pieces and sewed 'em all over the top of the gorilla's head, and that was it.

Today, many small-time movie makers are familiar with the methods Pierce used to make *The Legend of Boggy Creek*—"guerilla film making," as it has come to be known—but at the time Pierce was definitely taking a huge risk by investing time and other people's money into an independent venture that literally depended on the effectiveness of a modified gorilla mask and unproven actors. However, as time would tell, the movie would ultimately make a monkey's uncle out of any doubters.

THE LEGEND OF BOGGY CREEK

> If you're ever driving down in our country along about sundown, keep an eye on the dark woods as you cross the Sulphur River Bottoms. You may catch a glimpse of a huge, hairy creature watching you from the shadows.— From the narration of *The Legend of Boggy Creek*

Not only has *The Legend of Boggy Creek* become a B-movie cult classic, but it also has bragging rights for being one of the first horror films shot docu-drama style. Whether intentional or not, the film's gritty, piecemeal production and first-person accounts impart a sense of realism that makes the incredible story seem all the more believable. This technique, common today, was way ahead of its time and is a major reason why this cult gem endures despite its shortcomings.

Of the 60 or so "actors" eventually credited in the film, nearly all of them ended up being locals who played themselves, or stood in for other locals, to reconstruct the alleged encounters with the monster. Much of the film's soundtrack was overdubbed using a combination of suspenseful narration and eyewitness retellings (voiced in unabashed Southern drawl), which contributes to the overall feeling of authenticity. These people look and sound scared, as if Pierce had miraculously been there to film the encounter as it was actually happening. Because of this, it seems less like a movie and more like a horrifying live news broadcast. Not surprisingly, it was frightening to many viewers, especially during its original release in 1972. The movie scored a G rating, which drew in the younger audience, but the movie managed to spook the adults as well.

Admittedly, *The Legend of Boggy Creek* appears "dated" by today's standards, but it's not hard to imagine the impact it had in the drive-in back in the day. It was still a time of relative innocence, despite the on-going Vietnam war, so adults and kids alike were still susceptible to its scares. Sure, the lumbering gorilla-faced beast is less than convincing, but the story comes off as genuine. To anyone who had spent a little time driving around America's backroads, the possibility that something unknown might be lurking out there was a truly frightening notion… and to the locals, perhaps a frighteningly *real* notion.

Since it's a movie, people tend to think that much of the content was outright made-up or highly exaggerated by Pierce, but this is not the case. After extensive research, I have been able to correlate most of the scenes to something that was either reported in the newspaper or circulated by word of mouth. Of course some scenes were embellished for dramatic purposes — it is entertainment, after all — but nearly everything was based on real life reports.

The film opens with a sweeping view of the dreary, yet beautiful, Sulphur River Bottoms, an uninhabited area that's certainly large enough for a monster to hide. Replete with soaring birds, diving turtles, and a few beaver shots, the initial feel is more wildlife education than horror movie. But the mood is soon shattered by the wailing scream of a monster. As the scream fades, a young boy is shown running through tall grass; the boy, now grown to adulthood, provides commentary on his actions and motivation. Arriving at Willie Smith's gas station, he pleads with three old-timers to come look into "a hairy monster" his mother had seen in the woods near their farm. The men laugh it off, but after Smith mentions that this is not the first time the boy had come up there to make his plea, an immediate feeling of unease curls up in one's belly.

This scene was based on an incident that occurred in the 1960s in which a mother sent her seven-year-old boy on a two and a half mile run to the town of Fouke where he informed the landlord that they had seen a large hairy creature approach their home. Pierce does well to introduce the young boy (played by his son Chuck Pierce) in this way, so that later as the adult narrator he can reflect back on the story of the monster.

The next sequences provide an overview of Fouke and the surrounding area along with an introduction to J. E. "Smokey" Crabtree, who only appears briefly in the movie despite his considerable contribution. We are also introduced to Smokey's real life son, Travis. After the Crabtree meet n' greet, the intensity ramps up with the first creature encounter. In stereotypical Southern fashion, Willie Smith, portrayed in the movie as a doubting old-timer, unloads a double-barrel blast of buckshot at the monster who has come prowling around his house. The narrator concludes with haunting smugness: "Now he believes."

Next we are treated to more alleged real-life incidents acted out in pseudo-documentary style. These include the haunting encounters experienced by James and Fred Crabtree and Mary Beth Searcy in the Jonesville area prior to the 1970s flap. These were depicted accurately in the movie, including the events at the Searcy house in which she claimed that the monster stalked around their yard one evening after dark.

Another scene portrays what I believe to be a combination of two sightings, those reported by Lynn Crabtree and Kenneth Dyas in the mid-1960s. In both cases, the teenagers were out hunting when they came upon the Fouke Monster. Pierce did not attach a name to the boy in his film, presumably because he was not granted permission. It is well known that Lynn Crabtree did not want to participate in the movie, nor lend his name, and I must assume that Kenneth Dyas felt the same way. In cases like this, Pierce filmed a slightly modified reenactment without using names. Nonetheless, the story is fairly consistent with these boys' reports.

During these sequences Pierce demonstrates a masterful command of suspense, effectively juxtaposing claustrophobic scenes of frightened victims huddled in their houses with glimpses of the huge creature creeping out of the woods. Overall, Pierce does a commendable job of upholding the monster's integrity by showing him at a distance, behind trees, or in dark lighting; this proves to be a wise choice given the few times it is shown up close. Granted the creature FX do not come close to the eerie figure seen in the world-famous Patterson-Gimlin film shot a few years prior (in 1967), but compared to other Bigfoot-themed movies of that era, the monster of Boggy Creek is by far the most effective.

The next part of the film dramatizes Fouke's growing concern over the monster. After hearing of so many encounters, some of them described as threatening, the town decides to organize an all-out search. In this scene, a group of hunters and law officials set out on foot and horseback, accompanied by world-class tracking dogs, in an attempt to flush out the solitary brute. This is based on actual events in which local citizens and law enforcement officials organized searches. In fact, the dogs used in this scene were some of the same ones used in the actual hunts. In the movie, much to the dismay of their owners, the dogs refused to follow the rancid scent of the creature and instead tucked their tails and whined. This sequence has so much of a live news report feel that at times it's hard to remember that it's only an embellished recreation.

*Pierce demonstrates to Fouke resident and actor, Willie Smith,
how he wants to reenact the scene where Smith fired on the
creature from the porch of his home.
(Courtesy of the Texarkana Gazette)*

Descending darkness ends the search and segways into what is arguably the film's only major stumble. The narrator, theorizing that the monster is hurt, informs us that it retreated into the woods for a period of eight years. This time lapse is conveyed by more grand nature shots accompanied by a ballad that serves as the movie's main theme. The song melody has a down-home feel consistent with the rural setting, and it's catchy, but its overly sympathetic tone and lyrics seem out of place in a monster flick. Sadly, it breaks the mood and now dates the film.

In Pierce's defense, he needed some kind of interlude, and without a Hollywood budget he had to be resourceful. So he wrote the lyrics and sang the ballad himself! (The song is credited to "Chuck Bryant," a play on his own first and middle names.) At least he's in tune, right? When I spoke to his daughter, she assured me that her father lamented the fact that he did not have the money to invest in a better soundtrack. He just had to make do.

After a few verses, the movie snaps out of the love-in and returns to the tale, this time focusing on Travis Crabtree, who ventures off into the bottoms by himself on a trapping and fishing trip. Unfortunately, we are dealt another round of crooning at this point, a catchy-but-cheesy theme song that starts off "Hey Travis Crabtree / Wait a minute for me." One can only imagine that the song has haunted Travis in real life far more than the creature ever did!

During this sequence Travis drops by to visit an interesting man by the name of Herb Jones. Jones, who played himself in the film, was a real man who spent more than 20 years living in a remote shack deep within the Mercer Bayou near a place called Little Mound. It was difficult to scratch out a solitary existence in the unforgiving swamp, but to make things even more challenging, Jones had an unfortunate accident that left him disabled. Jones had been fishing in his boat one day when he accidentally shot himself in the leg. The gun was propped up in the boat and when he pulled up to the bank to get out, the gun slipped down and went off. With no one around, Jones was forced to crawl for several miles through the inhospitable terrain before he finally reached help. It was a miracle that he survived. In the movie Jones states that he has never seen the so-called Fouke Monster and makes it clear that he does not believe in such a thing. Jones has since died and his shack has been reclaimed by the bayou, although it is still possible to find Little Mound if you happen to know where it is.

After the movie takes a small detour with Travis, the monster finally returns with a vengeance. This time it is seen running across the road at dusk, scaring cows, and rustling around chicken coops. The road scene, although portrayed by teenagers, is presumably based on the real-life report by Mr. and Mrs. D.C. Woods Jr.; they saw the monster run across the highway while returning from Shreveport one night in 1971. The livestock scenes were based on conjecture by a few locals that the creature had been responsible for the unexplained deaths of their animals.

The next scene highlights the discovery of a series of strange, three-toed footprints in a freshly plowed field. This was based on the 1971 event in which Mr. Kennedy discovered the tracks in Willie Smith's soybean field near Boggy Creek. As in real life, local officials and wildlife experts are called in to investigate, but no one can determine what kind of animal (if it was animal) made the tracks.

Next, the creature frightens Bessie Smith and her children as it lurks on the edge of a field. This scene is based on an alleged sighting by Ms. Smith, although it is impossible to determine how accurately the encounter is portrayed. The following scene is short but no less effective, as it dramatizes the spooky eyewitness testimony of Charlie Walraven, who saw the creature one evening while driving near his home.

The monster's rampage continues as it terrorizes some high school girls who are having a slumber party sans boyfriends and parents. This scene, which was portrayed fairly accurately, was based on an incident that occurred in 1971 while Chris Rowton and two friends were alone one night in a trailer home near Boggy Creek. They heard noises all night long as something stalked around on the porch. The girls never saw anything, but the following day they found "large greasy tracks" left by some unknown animal. This sequence ends with a freeze frame on one of the girls as she screams in terror, providing one of the most chilling visuals in the entire film.

Pierce then unfolds his version of the hair-raising events reported by the families of Don Ford and Charles Taylor (referred to as "Charles Turner" in the film). As with some of the previous scenes, the men are away at night, leaving the women home alone. As if able to sense this situation, the monster begins to stalk the porch each night and even tries his hand at doorknob turning. When the men return home late in the night, they are told of the horrifying events and proceed to borrow a shotgun from the landlord. When the monster returns and thrusts his hairy digits through a window, they rush outside and commence "tah shootin'." Feeling that they've successfully injured or at least run the beast off, they return to the house for some sleep. But when younger brother Bobby Ford decides it's time to sit and contemplate the situation on the toilet—hey, if an ape-like monster came out of the woods, it would scare the crap out of you too!—the monster peers into the bathroom window as it returns for a final round. The ensuing confrontation culminates with Bobby being escorted by the police to a Texarkana hospital where he is treated for shock, just as he was in real life.

The tale is finally wrapped up by the little boy from the first scene—now an adult—as he returns to the old house where he and his mother had seen the creature so many years ago. As he looks out across the high grass, he wonders if the creature is still out there. Pierce lets the monster howl one last time as the sun sets.

As the monster's howl fades into the distance, we are left to consider these strange events. Although Pierce might have fudged a few facts, changed some names, and embellished the drama, it was all done in the service of entertainment, which is the prime objective of cinema.

Pierce takes the most liberties with the climactic tale of the Ford family, the longest scene in the movie. Though he sticks to the essentials of the report, he draws it out and adds details to create suspense. To set the story straight, I would like to present a firsthand account of the facts as told by Patricia Ford, wife of Don Ford. After a bit of digging, I located a letter she had written to the *Texarkana Gazette* after the movie's release in an attempt to sort fact from fiction. Her complete account is as follows:

> First off let me introduce myself. I am Mrs. Don Ford. But I'm writing this for three of us. Namely Don Ford, Bobby Ford and myself, Patricia C. Ford. I'm writing in regard to the article that was in the paper concerning Glen Carruth who played as Bobby Ford in "The Legend of Boggy Creek." You are welcome to print this if you so desire because I'm going to tell you the true facts of the night the Fouke monster visited us. In the movie it was mostly fiction instead of facts.

It was true about the two couples moving into the house. One couple was Don, our four children and myself. The other couple was Charles and Elizabeth Taylor. They had no children. At the time she was pregnant. The first night we stayed there we don't know who or what was fooling around the house. We did take the children and go across the highway to our neighbor's house, not the landlord's because he lived about 3 or 4 miles from us.

Bobby and Corky Hill did come to spend the weekend with us. They went to the creek to fish, but never did because they found the footprint[12] as soon as they got there. That night we had a visit from the Fouke monster. Don and Charles were not at work, they were there. The thing was either awfully smart or there were two of them because it would lead Don and Charles away from the house and before they got back it would already be there.

They did go to the constable's home after a gun and then he brought another one. The bathroom scene was true except Bobby ran into the living room instead of the bedroom.

When the hand came in the window, it was Elizabeth who was sitting there on the couch. Bobby was sitting in the chair holding a butcher knife. When he saw the hand, he grabbed Elizabeth and threw her on the floor.

The rest was true until the part where Bobby went all the way through the door. He didn't. He ran around the porch to the front door and just as I was about to open the door for him, he stuck his whole arm through the glass part of the door.

We did go back to the constable's house when Bobby was in shock but no one laid a hand on him except Don. Bobby kept coming to on the way to the hospital. He was screaming and trying to kick out the car windows so Don would knock him back every time he came to. We had no police escort either. We escorted ourselves to St. Michael's Hospital by blowing the car horn and running every red light from Fouke to St. Michael.

There you have the story as it happened. The Texarkana law did go out to the house with guns and look around but they couldn't find anything. It seemed to us that the thing was after our dog because every room our dog went into was where the thing tried to get in next.

Although Mrs. Ford's account may be more accurate in terms of the chronology of events, whether or not Bobby jumped through a screen door or shoved his hand through some glass, and whether they had a police escort or not, the fact is that the families experienced something very strange. They still believed a seven-foot hairy creature had visited their home on possibly more than one occasion, not to mention that they fired guns at the creature and *something* stuck a hairy hand or paw through the window. Despite any liberties Pierce took in the pursuit of entertainment, there is no denying the fact that something mysterious lurked in the culture, conscience, and surrounding countryside of down-home Arkansas. Now that something was about to experience its very own permanent cinematic legacy.

FOUR-WALL PHENOMENON

Once the raw movie was in the can, as they say in the business, Pierce packed it into his car and headed to Los Angeles. Mr. Ledwell had reached his monetary limit—roughly $100,000—at that point, so if the movie was going to be completed with editing and soundtrack, Pierce would need the help of someone willing to cut him a deal. He didn't have any Hollywood contacts, but after asking around, Pierce was put in touch with Jamie Mendoza-Nava, who owned a small post production house. Mendoza agreed to finish the movie without much cash up front, if Pierce would pay the fee for his editor and agree to give Mendoza a small percentage of the royalties.

Recognizing a fair deal—and his only option, really—Pierce turned the project over to Mendoza. The raw film was then edited by Tom Boutross (who had previously cut the classic horror film, *The*

Hideous Sun Demon) while Mendoza composed the soundtrack. In a short time, Pierce was the proud owner of a completed movie.

The next step was to find a suitable distributor. But, as expected, an independent filmmaker from Arkansas shopping a "Bigfoot" movie was not going to receive a red carpet welcome at any Hollywood studio. And, in fact, for a while it seemed he would not get so much as a bathroom rug to step on. He was repeatedly hung up on or laughed at by studio reps. Or he was told that he would need an agent to submit the film.

So Pierce needed another plan. Not only did he believe in his film, and was not a man to give up easily, but he owed Mr. Ledwell a rather substantial sum. Ledwell had been calling nearly every day to find out if he had any luck selling the film. Things were not looking good. So Pierce drove back to Texarkana and promptly headed to an abandoned, run-down theater. If nobody in Hollywood was interested, then he would just show the movie himself. In his *Fangoria* interview, Pierce said: "I went to the old Paramount Theater in downtown Texarkana. I went in, and I saw that the projectors and stuff were still there. I talked to an old projectionist, and he said, 'Oh yeah. They'll run.' So I called the company that owned the theater and asked if I could use it to have a world premiere. And they said, 'Well, Mr. Pierce, we'll sure *rent* you that theater.'"

Pierce convinced Ledwell to cough up $3,500 to rent the theater for one week and hire all the necessary crew including ticket-takers and a projectionist. This self-driven process was not common, but some, like Sunn Classics, had gone this route in order to play its low budget films. Known as "four-walling" (i.e., renting the entire theater to the "four walls"), Sunn had ironically done this a year earlier in order to market *Legend of Bigfoot*, a pseudo-documentary produced by famed cryptozoologist Ivan Marx. In fact, Pierce cites *Legend of Bigfoot* as having an influence on his own movie.

Since the theater hadn't been used for some time, Pierce and his crew had to literally hose down the floors and clean it up themselves. Nerves ran high. Would all their hard work pay off? Or would the film flop? But just one look around the block outside the theater on opening night was sufficient to answer that question. The buzz the Fouke Monster had created in the newspapers just a year earlier would have no problem in carrying over onto the big screen.

The movie premiered on August 18, 1972. Pierce sums up the excitement: "There were people lined up for four or five blocks. People had brown-bag *lunches* with 'em because they knew they couldn't get into the next showing, but they didn't want to lose their place in line. I knew it was gonna work when they started laughin' and getting excited, and screamin' whenever that booger'd jump out.... I got a G rating, and so the kids were in there screaming—it was scarin' the *devil* out of 'em. I knew then I had a winner."

In no time, Pierce pulled in enough cash to double his efforts. He took his only other print of the film, which was actually a reject copy, and began showing it in another rented theater in Shreveport, Louisiana. It was a hit there, too. Then came the calls from Hollywood studios, realizing they had bet on the wrong horse... er, monster. But they mostly wanted to "test" the movie in other markets without paying any money up front. Pierce wisely passed and continued to four-wall his own showings. It didn't take Pierce long to pay off Ledwell, Mendoza-Nava, and others who had kindly fronted the expenses, but so far he hadn't seen much return for his own investment of time and effort over the previous year. That changed dramatically, however, when he struck a deal with Howco International, a successful regional distributor.

Pierce asked for one million dollars (plus money to cover taxes) for a 50 percent interest in the movie. Howco balked at first but soon made Pierce's dream of becoming a millionaire a reality with one monstrous million dollar check. With a proper distributor, Pierce worked closely with Howco to promote the movie, transforming it from a four-wall phenomenon to a certified national sensation that played in theaters and drive-ins across America. Pierce was also able to cut a deal with American International Pictures (home of the low-budget cinema mogul, Roger Corman) to bring *The Legend of Boggy Creek*, and awareness of the Fouke Monster, to audiences worldwide.

The film was a remarkable success, going on to gross more than $25 million dollars, a figure backed up by notes in Pierce's own papers and verified by his daughter Amanda. Some claim the movie made upwards of $100 million, but $25 million is the official mark. It's certainly an amazing outcome for a first-time director who depended on volunteer help, civilian actors, and his own ingenuity to make it happen. And it was something the little town of Fouke could be proud of, right?

Wrong.

THE AFTERMATH

The timing could not have been better for *The Legend of Boggy Creek*. It was a time in the 1970s when interest in Bigfoot was on the rise and the media did not have such a cynical view of the unexplained as they have today. Despite the G rating, Pierce's docudrama style managed to not only scare the moviegoers, but to blur the line between the real and the sensational aspects of the subject matter. This resulted in fevered reactions ranging from real fright to all-out Fouke Monster mania.

Once again people began to descend on Fouke like wolves on a rabbit feast. The curious now came from *hundreds* of miles around, not just regionally, in hopes of either spotting the monster, having a look at houses shown in the movie, or to hunt the monster itself, as many locals had done a few years earlier. It was a flashback to the chaos incited by the radio station reward of 1971, but this time the craziness was even worse. As they did during the previous invasion, town officials forbade hunters to take guns into the woods to look for the monster except during deer season. It was all they could do.

New, larger bounties were posted. On September 16, 1973, the Texarkana-Arkansas Jaycees offered a whopping $10,000 reward for the capture of a live monster. The story ran in newspapers as far away as Victoria, a small town in south Texas. As expected, this drew even more hunters to Fouke, many of whom had shotguns perched in their truck's gun racks and dollar signs in their eyes.

Letters deluged the town, adding postal employees to the list of annoyed locals. At one point the Fouke Mayor's office had a bundle of nearly 800 letters asking about monster hunting. One letter read: "We have seen a movie… and would like some information on the Boggy Creek monster as well as some information on the country you are in." Another added a sense of urgency: "We are interested in receiving all information that you could possibly send us concerning the creature there in your area. We feel that this is very important to mankind." Some letters were addressed to the Fouke Monster itself!

With the new wave of tourists, a few townsfolk recognized the opportunity and unveiled some sophisticated merchandising. Bill Williams, the new owner of the Boggy Creek Café, added several items to the menu such as the "Three-Toed Sandwich" and the "Boggy Creek Breakfast." Willie Smith went further by manufacturing an unbelievable selection of money clips, cards, key chains, ash trays, bumper stickers, license plates, and other items engraved with "Home of the Fouke Monster." He sold these out of the café and his gas station next door. Smith's son, Monroe, later remarked that he was getting more questions from tourists about the monster than gas sales.

Smith also whipped up a new batch of his famous three-toed casts. These casts are now considered collector's items in the cryptozoology community, since the originals were lost forever when the café burned down in the late 1970s. If anyone still has a first generation copy of the cast, especially if it's signed by Willie Smith and Smokey Crabtree, you've got a genuine treasure on your hands!

Jane Roberts, the wife of future mayor Virgil Roberts, also got in on the hype by creating her own miniature footprint replicas. She reportedly manufactured 5,000 copies, complete with a hand-painted "Greetings From Boggy Creek" message, and sold them to outside distributors for 50 cents each. A 45-RPM vinyl record was released, featuring a song called "Fouke Monster" by Billy Cole and The Fouke Monsters. The song, commissioned by Pierce as movie advertising, sounded like quirky 60s-era rock with a spooky chorus that chanted "Monster, Fouke, Monster, Fouke, Monster."

7-Inch 45 RPM vinyl record by Billy Cole and The Fouke Monsters released in 1971.
(From the personal collection of Lyle Blackburn)

The little town of Fouke (population 509 at the time) was not ready for the new wave of madness that came their way as a result of the movie's success. Fouke Mayor J.D. Larey summed up the surprising outcome in an interview with the *Texarkana Gazette*: "The man who made the movie had never made movie in his life. The guy who backed the movie had never backed a movie in his life. The people who acted in the movie had never acted before. I don't think you could have foreseen anything happening on it."

At this point, just about any animal with hair or legs seen skulking around Fouke was likely to be fingered as a monster.

However, a few fairly credible reports still managed to stand out against this background noise. On November 25, 1973, Orville Scoggins watched a black-haired, four-foot tall creature creep across his bean field one morning. "I looked up to see what the noise was about, and there it was, about 100 yards from where I was, walking eastward very slowly," Scoggins told reporters from the *Texarkana Gazette*. He confirmed that the animal "stood upright on two feet" and estimated its weight at "80 or 90 pounds."

After the sighting, Scoggins jumped in his pickup and raced down to the Fouke Café where he found Constable Red Walraven. Upon hearing the story, Walraven and two other men went back to Scoggin's farm, which was located on County Road 9 just four miles from Fouke's main strip. Upon inspecting the area where the "monster" had been sighted, they found a line of visible tracks left in the soil. The tracks measured 5.5 inches in diameter and were spaced 40 inches apart. The men traced the tracks for nearly an eighth of a mile before the trail finally disappeared into the woods. An immediate search of the area, unfortunately, did not locate the culprit.

These tracks were shorter than the 13 inch tracks found in Willie Smith's field, but Walraven noted, "These were the same tracks that were found the first time the monster was sighted." (Presumably this meant they were of the three-toed sort.) In the end Scoggins was not confident that he had seen the Fouke Monster, but he was positive that he had seen *something*. Scoggins had always been known as a die-hard monster skeptic, but after this incident he changed his mind saying, "something stalks the woods near Fouke." I spoke to Lloyd Sutton, another long time local resident, who confirmed the details of this story and reiterated that Scoggins had always been a non-believer and was not a man to make up stories.

Another incident, which occurred around 1974, came to me by way of Doyle Holmes, who grew up in the remote area along County Road 35 south of Mercer Bayou. Holmes was about nine years old at the time, but he will never forget the night his neighbors, the Giles brothers, came roaring into his house looking for his father. They had been out deer hunting that evening and had bagged a

1974: Two brothers report an encounter with an angry creature south of Mercer Bayou.

The old Giles home, now collapsed, near the site of the "ape attack."
(Photo by the author)

big buck. But they were not coming to brag about the kill. Instead, they were in a state of sheer panic. They claimed that as they were driving their truck back up the small road that led to their house, a hairy, ape-like creature suddenly emerged from the woods and jumped into the back of the trailer they were towing at the time. It proceeded to thrash about, damaging two motorcycles strapped to the trailer's bed. When he realized what was happening, the driver hit the brakes, but the two brothers didn't dare get out of the cab. The creature looked enraged as it took its fury out on the hapless bikes. After several seconds and considerable damage, it jumped out of the trailer and ran back into the woods.

Holmes does not remember much about their description of the creature or other details, only that the men seemed genuinely upset by what had happened. His father and the men headed off to investigate, but they never found any sign of the angry beast.

The Fouke Monster has reportedly displayed aggressive tendencies, but jumping onto a moving vehicle and fighting motorcycles seems a bit far-fetched. At the time, everyone in the area was certainly familiar with the legend, so it's possible that it was some kind of joke, but at that young age Holmes could not be sure. Regardless, we visited the Giles property one afternoon out of curiosity. The road where the incident allegedly occurred had long since grown over with grass, and the original house is now collapsed. Interestingly, the isolated property is only a short distance from where Pierce filmed the dog tracking scene in *The Legend of Boggy Creek*.

While much of the monster ruckus was a headache for law enforcement, or just plain nonsense to others, it did impact some residents in more serious ways. Families who had property located on the outskirts of town had to contend with growing waves of sightseers and monster hunters trespassing on their lands, and at times, causing considerable property damage to fences or other structures. The folks with farms had to deal with people trampling through their crops, which could impact their already fragile incomes. The monetary gains brought in by the throng of wild tourists were not enough to balance out the fact that things were being destroyed in Fouke as a direct result of the monster's new fame.

Chris Rowton, a young girl in Fouke at the time whose story was reenacted for the movie, told me that people had little regard for the private properties near Boggy Creek. "There were people who would come searching, and they would camp out on my mom and dad's front yard. They would sit out there and watch for stuff or just listen," she said, thinking back to the craziness. "There for a while, you never knew who was going to come from where!"

While no doubt many locals experienced property damage and woe, their tales did not become public record like those of Smokey Crabtree, who seemed especially impacted by the movie-spawned madness. After all of Smokey's experience with the monster—his son's encounter, his relative's sightings, his now regrettable involvement in the Pierce film—he decided to tell his side of the story by way of a memoir. The result was a modest-sized book entitled *Smokey and the Fouke Monster*, which he self-published in 1974. The book offered a personal perspective of the events that amounted to nearly a decade of woe—all because of an undocumented animal that had eluded capture.

For much of the book, Smokey relates stories of his upbringing in the swampy backwoods as a way to underscore his honest and straight-forward viewpoints. Tales of catfish noodling, hog hunts, and close encounters with snakes are told to great effect in simple, country boy language. Eventually, he brings in stories of his early adulthood—as a boxer, navy man, pipe welder—and then finally covers his family's experiences with the Fouke Monster. While it is essentially a book about Smokey's life and experiences, the monster does play a significant role in the book, and as such serves as one of the only written records of several interesting Fouke monster incidents.

The primary motivation for writing the book came as a direct result of Smokey's involvement with *The Legend of Boggy Creek* film. After providing guide and consulting services to Charles Pierce, Smokey felt slighted by the film's astounding success, which left Pierce a millionaire and Smokey with a rash of trespassers trampling his land. Smokey had also been instrumental in convincing many of the locals to share their stories with Pierce, to act in the film, and in some cases loan out their residences as movie sets. Pierce had very little budget at the time, so—according to Smokey—Pierce promised most of the folks some payback once the film was purchased and released down the road. But once the premiere had come and gone, no one received a dime. Many of the Fouke locals had been invited to the movie's premiere, but this only led to more problems when they saw themselves on the big screen acting out a scene while the voice-over they heard belonged to someone else. Of course many of the townsfolk were just happy to see themselves in a movie, but still some were angry. Feeling taken advantage of, they turned to Smokey for answers. Smokey had been in contact with Pierce all along, and they expected him to iron out a contract for everybody before the movie went into circulation, but that had not been the case. Not surprisingly, the unfinished monetary business left an ugly scar on the relationship of the two men and ultimately pitted them against each other in a Los Angeles courtroom in the late 1970s.

But the more immediate concern was the horde of monster hunters beating a path toward the Crabtree property. After having their stories sensationalized for profit by others, and feeling Smokey had let them down, some of the townsfolk were more than happy to direct tourists straight over to Jonesville where they could get a glimpse of the Crabtree boys. This was not the first time Smokey had dealt with monster hunters, but it was certainly the worst. The first round had come in 1965 when his son encountered the monster at their private lake. Well-meaning locals dropped by to offer a hand with the hunt, which went well enough, but nosey outsiders also dropped in to hunt on their own. This resulted in a loss of privacy, which was important to Smokey and his family, and some property damage as people jumped fences and tromped across his land. But this was mere child's play compared to the foolishness that came as a direct result of the movie's success. Though nobody could have predicted the reaction, least of all Pierce, it did not make the bad situation any better for the Crabtrees. According to Smokey, in his first memoir: "They had seen the movie, *The Legend of Boggy Creek*, and they were taking my place apart looking for the Monster."

Some of the visitors were even insulting, calling the Crabtrees liars and phoning them at all hours of the night to ask silly or downright rude questions about the monster. It became a full time job to police the property and to keep the damage to a minimum. "They were coming by the car load and by truck loads. Sometimes, there was as many as twenty different groups here at the same time. All of them wanted something from our place to take home with them. Some were willing to settle for a twig from a tree they remembered seeing in the picture. Some asked for bark off the trees."

Smokey called Pierce and asked for help, but the mobs were out of Pierce's control. It seemed the monster's legend was blowing up like a puffer fish, and getting just as ugly. Smokey tried posting signs which read "PRIVATE: Monster Hunters or Sight Seekers Are Not Welcome" and "Lake Closed—Sickness." This helped for a while but ultimately did not keep people out. According to Smokey, a group of brazen sightseers simply tore the signs down, apparently mad that they had driven so far and were not able to speak to Smokey in person. But their wish was soon granted, in a backhanded sort of way, as Smokey chased them down with a shotgun. They all ended up at the sheriff's office.

It seems nearly inconceivable that a low budget movie featuring a rather cheesy long-haired "gorilla" would create such a stir, but behind it all there lurked a harmless monster mystery and that was part of its appeal. People tired of the horrors that had become so familiar in the bloody trenches of Vietnam were hungry for the escape provided by hometown movie monsters. The Fouke Monster—scary, mysterious, and perhaps real—was something that would thrill them but not kill them. It was aggressive at times but not a true threat. Sure, maybe he had caused Bobby Ford a few scratches and a trip to the hospital, but so far it had all been in fun. Fun for the whole country—with the exception of a small town called Fouke, of course. The front line is not usually the best place to be in a skirmish.

CREATURE FROM BLACK LAKE

The success generated by *The Legend of Boggy Creek* was not lost on other filmmakers, most notably a father and son team out of Shreveport, Louisiana, by the name of McCullough. The team consisted of writer/actor Jim McCullough, Jr., and his father, Jim McCullough, Sr., who had co-produced a 1974 film adaptation of *Where the Red Fern Grows*. Influenced by the Fouke Monster and other hairy cryptids such as the "Caddo Critter"—another large, ape-like creature said to inhabit the Caddo Lake area—the McCulloughs set out to create their own entry, a film known as *Creature From Black Lake*.

One-Sheet Poster painted by Ralph McQuarrie.

Caddo Lake with cypress trees and Spanish moss.
(Photo by the author)

A standout among the classic "Bigfoot horrors," the film was written by McCullough Jr. and made possible by his father's company, the aptly named Jim McCullough Productions. Like Pierce, McCullough teamed up with low-budget distributor Howco International Pictures, and even went so far as to bring in Joy N. Houck, Jr.—son of Howco's president, Joy Houck, Sr.—as director. This double dose of nepotism might seem like little more than a ploy to keep a few untalented offspring busy, but such is not the case. Both Houck, Jr., and McCullough, Jr., do a fine job creating a memorable film that further contributed to the reputation of Southern swamp monsters.

Although *Creature From Black Lake* is a much more conventional movie than *The Legend of Boggy Creek* in that it doesn't use a narrated docudrama approach, it still makes excellent use of local legends and real life set locations that can be linked to similar "monster" sightings. While McCullough Jr.'s script is purely fictional, local legends of mysterious creatures, including those of the Fouke Monster, must have served as inspiration.

The story is set in the real town of Oil City, Louisiana, which is roughly 120 miles south of Fouke near the east side of Caddo Lake. Like Fouke, Oil City is basically a one-horse town located near a suitable swamp where hairy monsters are seen on occasion. The movie's title was presumably taken from Black Bayou Lake, a smaller body of water located to the north, although principle filming of the swamp scenes was actually done in the hauntingly picturesque setting of Caddo Lake.

The unique features of Caddo Lake make it an ideal location for a monster flick, and no doubt *Creature From Black Lake* benefits greatly from its spooky beauty. Everywhere you look, sinewy cypress trees rise from the water and stand like grim sentries around the shoreline. Each one is draped in a gray shawl of Spanish moss to give it an ancient and menacing ambiance. Like Mercer Bayou in Arkansas where Pierce captured amazing shots for his movie, the internationally protected wetlands surrounding Caddo offered a similar backdrop of monstrous style for Houck and McCullough.

According to the Texas Bigfoot Research Conservancy, one of the regional groups dedicated to the research and discovery of the animal known as Sasquatch, the areas surrounding the lake have been the scene of dozens of Bigfoot sightings since 1965. So it would seem that art is imitating life in *Creature From Black Lake*, or perhaps it's the other way around. Whatever the case, the location seems to lend a bit of authenticity to the film, ultimately resulting in an entertaining ride despite the film's low budget.

The success of *Creature From Black Lake* was nowhere near that of Pierce's seminal effort, but nonetheless it stands today as a good entry into the Bigfoot film cannon and in some ways helped stoke the fire started by the Fouke Monster five years earlier.

Lasting Influence

The phenomenon generated by Pierce's homemade film has rarely been repeated in the history of celluloid horror. In fact, only a handful of these DIY movies have been able to equal the level of fear feedback and box office returns enjoyed by *The Legend of Boggy Creek* during its prime. The two best examples are *Blair Witch Project* (1999) and *Paranormal Activity* (2007), both of which managed to achieve major success despite their lowbrow production values.

Blair Witch Project was actually heavily inspired by the *Boggy Creek* movie. In a 1999 interview with *The Tulsa World*, *Blair Witch Project* co-director Daniel Myrick stated: "We just wanted to make a movie that tapped into the primal fear generated by the fact-or-fiction format, like [The] Legend of Boggy Creek."

I can remember the buzz when *Blair Witch Project* hit the theaters. Like *The Legend of Boggy Creek*, moviegoers were quick to buy into the realism conveyed by the shaky, handheld camera and improv scripting. It was the early days of reality television and the point-of-view camera channeled the character's fear straight from the screen to the audience with fantastic results. The "monster" of the film (the presumed ghost of the Blair Witch) is even more shadowy and elusive than the one that stalks Boggy Creek, allowing it to creep more effectively into the sophisticated psyche of modern horror fans. Like Pierce, the directors managed to blur the lines between reality and cinema, resulting in a huge box office draw and ultimate success.

More modern ghostbuster creep-outs such as *Paranormal Activity* (2007) and its sequels use a similar docudrama style, stripping it down even more to a level of stark reality. Made on a shoe-string budget and filmed with a home video camera, *Paranormal Activity* explores the universal fear of ghosts, house hauntings, and the vulnerability one can experience while sleeping. Once again, the result was an unanticipated level of success with worldwide distribution and sequel deals reminiscent of *The Legend of Boggy Creek*. And the similarities don't stop there. Just as *The Legend of Boggy Creek* incited monster hunters to take to the woods with guns, this modern ghostly counterpart helped push a wave of ghost hunters into old buildings with EMF detectors, proving that real life creepies often strike the most resonate chord within our primal psyche.

Another homegrown indie that pays homage to Pierce is the 2009 film, *The Wild Man of the Navidad*. Directed by two Austin-based filmmakers, Duane Graves and Justin Meeks, this tale evokes *The Legend of Boggy Creek* in both style and subject matter.

"*Boggy Creek* was a huge influence when we were developing *The Wild Man of the Navidad*," director Duane Graves told me as we discussed our mutual love of the film. "We wanted to do the same for our part of Texas as it did for Fouke, and make the Wild Man a character of the movie as much as the region itself. That's the main reason we decided to use real life residents of the area…"

Watching *The Wild Man of the Navidad*, it's clear that the directors meant it to be a tribute to Pierce. "Several visual homages and verbal nods to *Boggy Creek* appear in the finished product," said Graves. "It was definitely something we wanted to display proudly when making the film. We even spoke with Pierce several times in hopes of bringing him aboard as a producer of the project, but he was more interested in directing it." According to co-director Justin Meeks, they talked to Pierce at least a dozen times on the phone. "He was thrilled we were continuing to spread the word about the possibility of a Sasquatch," Meeks told me. One wonders how Pierce might have contributed to the film, but the young directors do a fine job with their own Texas legend.

It is worth noting that Pierce came to believe in the possibility that Sasquatch-like creatures, such as the Fouke Monster, may exist. Not only is this point conveyed by Meeks' conversations with the director, but more directly, Pierce's daughter told me that, based on comments he had made to his family and others, she had no doubt he believed in the creature.

A final interesting note regarding the legacy of *The Legend of Boggy Creek* is associated with its eye-catching movie poster. The

One-Sheet Poster painted by Ralph McQuarrie.

simple, yet striking, image of the creature as it is coming across the swamp has become the iconic image of the Fouke Monster. It was painted by none other than Ralph McQuarrie. Though not a household name, McQuarrie's fantastic imagery is known worldwide. McQuarrie also painted the posters for many of Pierce's subsequent movies, including *Boggy Creek II*, *Bootleggers*, *The Town That Dreaded Sundown*, *Winds of Autumn*, and *Winterhawk*. And he created the memorable poster for *Creature From Black Lake*. These are all great works, but they are not McQuarrie's most notable achievements. What most people don't realize is that McQuarrie designed many of George Lucas' classic Star Wars characters, including Darth Vader, Chewbacca, Yoda, Boba Fett, the stormtroopers, and others. He also painted various concepts for the film's sets, including R2-D2 and C3PO's arrival on Tatooine, the Jawas, and many of the Deathstar sequences. McQuarrie's fantastic design work, in fact, helped convince 20th Century Fox to initially fund Star Wars.

I had the good fortune of conducting an interview with McQuarrie in regard to his work with Pierce. He told me that they were introduced by a mutual acquaintance during the final phases of the film. Although McQuarrie had never done a movie poster before, he was already a successful graphic illustrator. As a graphic designer himself, Pierce recognized McQuarrie's talents and left the job in his capable hands. "He came over to my studio and we talked about the movie," McQuarrie told me. "I came up with the design of the movie poster, he liked it and that was that."

In contrast to many of the exaggerated and outright misleading horror movie posters of the 1950s and '60s, *The Legend of Boggy Creek* poster employs a more understated approach, portraying the monster as more of a shadowy figure, just as Pierce does in the actual film. The use of backlighting, provided by the setting sun, allows the monster's features to remain in shadows, yet still shows his full ominous form as he moves through the swamp. The placement of the sun, as it is just beginning to drop behind the trees, also implies that nightfall will soon be coming… and with it, the monster!

The mastery of these subtle elements and the overall eye-catching design of the painting showcased McQuarrie's potential for movie posters right from the start. Due in part to this early work with Pierce, his body of work went on to include concept and poster designs for many blockbuster films, including *Battlestar Galactica*, *Close Encounters of the Third Kind*, *Star Trek*, *Back To The Future*, and many others.

After learning of the connection between *The Legend of Boggy Creek* and *Star Wars*, naturally I wondered if somehow the Fouke Monster had managed to influence the character of Chewbacca, who has some definite Bigfoot-like traits. But after speaking with McQuarrie, I had to begrudgingly let go of the idea that the Fouke Monster had played any such role in sci-fi history. Any resemblance between the creatures for Boggy Creek and Chewbacca, McQuarrie told me, was purely coincidental. "George Lucas had a very specific idea for Chewbacca," McQuarrie explained. "I worked from his description."

In my final question to McQuarrie, I asked how he liked working with Pierce on the various posters. "I did enjoy working on those projects," he said. "I was given carte blanche when it came to the design, and Charles was pleased with the final outcome."

Pierce had good reason to be pleased with *The Legend of Boggy Creek* movie poster. It stands as a classic image even to this day.

5. It's Still Out There

Legacy Makers and Caretakers

Following the movie-inspired madness, things began to calm down for the town of Fouke... and the monster as well. The film had run its course on the drive-in and theater circuits, and the newspaper stories had become nothing more than rehashed vignettes of the original stories dragged out of the closet for the Halloween season. It seemed that the movie had made the monster famous but at the same time overshadowed his own possible reality. With no body or other proof, the Fouke Monster—or as it was now referred to by the general public, the Boggy Creek Monster—became more associated with his movie persona than with the possibility that there just might be a family of unknown species inhabiting the swamplands of southern Arkansas. Sure the movie scenes were played for maximum effect, but this did not erase the fact that they were based on allegedly real events. However, in the wake of the movie the general public's fascination moved on, leaving the burden of proof on the shoulders of the monster once again.

Despite the diminished public interest, however, visits from dedicated monster enthusiasts did not stop, nor did the sightings. Credible reports were still turning up, although the news media was somewhat more reluctant to follow up on them than they had been in the past. So unless you were a Fouke local, it would have been tough to follow or even hear about the sightings during the pre-internet years. Fortunately, a few Fouke residents still took the monster seriously and did what they could to preserve his legend and further his credibility.

Two of the most prominent champions over the years have been Smokey Crabtree and Rick Roberts. Both men were significantly

Newspaper ad using the Fouke Monster to sell Halloween costumes.
(Courtesy of the Texarkana Gazette)

involved with the mystery since before it was made famous by the movie, and they remained interested after Hollywood had packed up and moved on. Eventually, by virtue their own industrious natures, both men would put themselves in a position to become a caretaker, so to speak, of all things Fouke Monster.

Another significant player with regards to Fouke Monster history is Frank McFerrin, the curator of a small museum in Fouke and author of two local history books. His willingness to include the legacy of the Fouke Monster in his endeavors has further served to solidify the creature's place in Miller County history.

I've spoken to all three of these gentlemen on numerous occasions, sometimes at length, and their input has been invaluable. Smokey, of course, is so intertwined with the Fouke Monster's history that it would be impossible to tell the complete story without him. In the case of McFerrin and Roberts, both have been instrumental in relaying old and new stories, and in helping me track down other knowledgeable locals. So I feel it is important to briefly introduce these men and highlight their contributions to Fouke Monster history.

J.E. "Smokey" Crabtree

As we have already seen, Smokey played a huge role in the monster's history by way of his family's early sightings and by acting as consultant on the original movie. By 1974, when he published his first autobiographical book, *Smokey and The Fouke Monster,* his association with the monster had already been firmly established. Throughout the years, he has continued to seek proof of the creature, and to this day he remains involved with its ongoing saga.

I have spent time with Smokey on many occasions starting with our first meeting in 2009 at a Bigfoot-related conference in Tyler, Texas, that featured Sasquatch researchers, enthusiasts, and witnesses. I was able to purchase Smokey's first book, which he graciously signed, while I told him of my affection for *The Legend of Boggy Creek* film. Since then, I have visited his bookstore in Fouke where we discussed a variety of subjects, and I saw his interesting collection of hunting relics, newspaper clippings, and Fouke Monster related items. Smokey's bookstore is located on his personal property but is open to the public as his time permits.

I found Smokey to be a straight shooter, full of amazing stories and observations about growing up in the Arkansas countryside, as well as about the creature. His no-nonsense sincerity comes across both in person and in his books, which is another reason they make for a fascinating read. And even though the monster has been somewhat of a blessing and curse to Smokey and his family over the years, he continues to speak about the subject and, whenever possible, assists others in an effort to solve the mystery.

The irony of it is that while Smokey has been forthcoming as a writer, recording details about the monster's legend that would have otherwise been lost in time, the "monster" has never afforded Smokey the chance at even the briefest of sightings.

Frank McFerrin

McFerrin is the long-time curator of a small museum in Fouke that's operated by the Miller County Historical Society. The museum is tucked away on an inconspicuous side street, which meanders off the main thoroughfare. The quaint building houses a wide array of historical items that illustrate the life and times of the hard working people of Miller County. It also features a section dedicated to the Fouke Monster, which includes photos, news clippings, memorabilia, and an original *The Legend of Boggy Creek* movie poster, which is prominently displayed in a handcrafted wooden frame. Locals often gather at the museum to talk about current news or reminisce about days gone by. Inevitably, talk of the Fouke Monster pops up from time to time, and I've been fortunate to hear many of their recollections and tales about the creature while in the laidback atmosphere of the little museum.

McFerrin's two books about Fouke and the surrounding area provide an in-depth look at life in Miller County through words and pictures. Having lived in Fouke all of his life, McFerrin knows most of the locals who claim to have seen the creature, and respects the Fouke Monster—real or not—as a valued part of local history and culture.

McFerrin has been invaluable to my research, providing photos and information as needed. During my numerous visits, I found him to be an extremely knowledgeable and down-to-earth person, who pays careful attention to factual details.

Rick "Rickie" Roberts

Roberts is another Fouke local who has spent much of his life in the shadow of the monster. His father and mother have both served as the Mayor of Fouke at one time or another, and his brother Denny owned the Boggy Creek Country Store until 1992. As a teenager, he was not able to directly participate in the movie, but he was on site when authorities examined the mysterious three-toed tracks. As an adult, Roberts played a more high profile role as owner of the Monster Mart convenience store located on Fouke's main strip. Being the only convenience store in town for many years, as well as the only business that still incorporated the monster into its marketing, the Monster Mart became a natural depository for

*Rick Roberts in front of the original Monster Mart mural painted by the
late Karen Crabtree.
(Photo by Chris Buntenbah)*

tales about the monster. Many sighting accounts, both old and new, were often passed around in casual conversation at the store. Since there was no official place to make a report, other than the Sheriff's office, it only made sense to drop by the Monster Mart and pass on the information to anyone present, usually Roberts himself.

The Monster Mart was also a regular stop for the media, when they came to town for a follow-up story about the monster. In the process, Roberts became something of an unofficial spokesperson, frequently going on camera to recap the history or to provide details about the latest sightings.

Roberts no longer owns the store, but still he has been instrumental in helping me collect and verify information for this book in an effort to finally document the monster's complete history. He and I have spent many hours driving around the

Fouke area, visiting locations where the monster has been seen and discussing relevant facts. After hours of casual interview and countless electronic messages, I learned not only about his personal experiences with the monster but those he heard of from others.

Roberts has a passion for the Fouke Monster rivaling that of anyone I've ever met. He expressed this passion through his ownership of the Monster Mart, where he initially constructed the small showcase of monster relics and newspaper clippings. He also commissioned the murals for the outside wall, doing his best to preserve all aspects of the legend.

Roberts has had a lifelong desire to find proof of the mysterious creature. Over the years he's heard several strange howls down in the bayous that he couldn't explain, but that's as close as he's ever come to the monster. That's too bad, since not only would his personal curiosity be satisfied, but there might be a reward to boot. During his tenure at the Monster Mart, representatives of a certain well-known cable television channel dropped by to make a few offers. According to Roberts, they offered $100,000 for a verified photo of the creature. If he happened to bring in a dead specimen, the reward would increase to one million dollars. And if he could manage to capture a live one…. a staggering *five million!*

RETURN TO BOGGY CREEK

By the last half of the 1970s, the country had weathered the storm of Vietnam and was now gearing up for the bicentennial and the coming of John Travolta. Prosperity and progress were spreading like bellbottoms. The interest in Bigfoot creatures – initially kicked off by the Patterson-Gimlin film of 1967 in which Roger Patterson and Bob Gimlin captured an alleged Sasquatch strolling along a dry sandbar in Bluff Creek, California—had been fueled by several Hollywood renditions, including the horrendous *Bigfoot* in 1970 and, of course, *The Legend of Boggy Creek* in 1972. The country's 200th birthday would also be marked by a flood of Bigfoot Americana, starting with *Creature From Black Lake*, followed by the feature film/documentary hybrid *The Legend of Bigfoot*, and finally by *Blood Beast of Monster Mountain*, all released in 1976. Bigfoot also made tracks on the small screen with guest appearances on the popular prime time television show, "The Six Million Dollar Man" (as a robot monster, no less), and a Saturday morning kid's series called "Bigfoot and Wildboy" in 1977. These portrayals tended to paint Bigfoot as something of a spectacle, but nonetheless kept the interest in the unexplained phenomenon at an all-time high.

More Bigfoot feature films were released in 1977, including *Sasquatch: The Legend of Bigfoot*, as well as *Return To Boggy Creek*. Yes, even back in those days a successful film would spawn an inevitable slew of sequels in an attempt to suckle more money from the public. But in the case of *Return To Boggy Creek*, it didn't do much except suck. The studio had been urging Charles Pierce to do a sequel, but as he tactfully put it in the *Fangoria* interview: "I was still trying to prove myself as a filmmaker; I didn't want to have to turn around and shoot the same thing all over again."

Figuring that the subject matter alone was enough to sell the film, the studio decided to move ahead without Pierce. They enlisted Tom Moore as director, a relative newcomer who had directed one prior horror movie at the time called *Mark of the Witch*. Considering its budget, Moore had done a fairly respectable job with that film but the deck was stacked high against him when it came to the Fouke Monster sequel. Duplicating the success of *The Legend of Boggy Creek* would have been challenging to anyone, and, as expected, even Moore could not pull it off. The resulting film was an embarrassing flop.

To begin with, the sequel completely abandoned the frightening pseudo documentary style that had worked so well for Pierce. Instead, the producers went with a more traditional horror movie approach, incorporating the television star talents of Dawn Wells (*Gilligan's Island*) and Dana Plato (*Different Strokes*) to carry it. The cheese factor of its stars certainly does not help the film, though they do a superior job of acting compared to the other cast members. In fact, the novice acting by the Fouke locals in Pierce's film outshines most of the acting in *Return To Boggy Creek*. Further contributing to the film's downfall, the real voices featured in the first film, with their authentic Arkansas drawl, are noticeably missing from the sequel. Instead we are forced to endure fake Cajun accents delivered by unconvincing actors.

All of those factors aside, the main problem is that it tries to be a heartwarming Walt Disney film in the vein of *Escape To Witch Mountain*. The plot centers around three children who are eventually stranded on Boggy Creek in a storm, only to be rescued by the monster himself. Like most of the big-screen Disney films of the era, the children are the main characters, and through their eyes the monster is transformed from a savage man-killer to a loveable brute, all outside the comprehension of the worthless adults.

The most puzzling element is why the filmmakers changed the name of the Fouke Monster to "Big Bay Tye." They still refer to the waterway where he lives as Boggy Creek, but there is no mention of Fouke anywhere in the movie. Instead, the setting is moved to a place called Happy Camp, which seems appropriate since the movie is devoid of any of the spookiness or realism that made *The Legend of Boggy Creek* so frightening six years earlier.

The Night Walker

As the popularity of California's Bigfoot grew and *Return To Boggy Creek* did little in the way of actual return, the Southern Sasquatch legend was mostly left to its quiet existence back in the swamplands. Reports of the creature are difficult to find for the remaining years of the 1970s and early 1980s, though several interesting sightings started to emerge from discussions with the locals who were still living in Fouke at the time.

The first sighting is quite possibly one of the most credible and amazing eyewitness accounts of the Fouke Monster that I have come across. I heard the details of the story directly from the eyewitness, Terry Sutton, and from his father, Lloyd Sutton, who was there to see the fear and shock on his son's face just after the encounter.

The date was February 20, 1981. It was an unseasonably warm afternoon. Fifteen-year-old Terry Sutton had collected a jarful of nightcrawlers from his mother's garden and headed off with his fishing gear towards the pond on the far side of their property. Their home was located in Jonesville, just off the main road that connects their countryside community with the outside world. Having grown up in the Jonesville/Fouke area, Terry was no stranger to the lurid tales of their haunting hair-covered beast, but this was not something that crossed his mind very often when traversing the backwoods. Like most boys in the area, Terry was an experienced hunter, trapper, and fisherman who had spent countless hours in the rich bayous. He was very familiar with all varieties of local wildlife and would not likely mistake one for a seven-foot-tall hairy hominoid. Perhaps that's why this event would be so shocking to him.

After traversing the quarter mile from his home to the pond, Terry loaded his gear into the small aluminum boat that his dad kept there and pushed off into the lazy water. He promptly baited a hook with one of the fat worms, dropped it into the water, and sat there quietly fishing for the rest of the afternoon.

But while sitting there enjoying the solitude, he heard something moving through the thick blanket of leaves that covered the late winter ground. Earlier, he had heard what he thought was the bellowing of his uncle's old Black Angus bull who roamed the adjacent property, so naturally he assumed the bull had wandered toward the pond. That, or perhaps his father was coming down to check on him. Either way, the sound was nothing out of the ordinary, since anything walking through the leaves would have made a significant amount of noise.

About an hour before dusk, Terry decided to paddle the boat around a small bend in the pond where the fishing might be better. The bend was like a small neck of water that jutted off into a wooded area. He heard some more loud rustling in the leaves coming from that direction, but still he wasn't alarmed... until he rounded the bend. Now he could see the source of the leafy noise. It was a large, hair-covered animal walking on two legs away from the pond. Its back was facing Terry as it headed toward a ravine that dipped down to a nearby creek.

"It was walking away in a casual stroll," Terry told me. "At first, I couldn't believe what I was seeing."

Terry was a mere 60 feet from the creature, so there was no mistake that he was seeing something other than human, bear, or any other common animal. Terry was over six feet tall at the time, so he estimated the creature's height to be as tall or taller than he was. He described its fur as being scraggly, three-to-five inches long, and colored a dull black or very dark brown. It had notably long arms, which swung as it walked, giving it an ape-like quality, although it did not hunch over as an ape might do. "I know there's varying stories, but what I saw was not bent over. It was walking upright just as straight as I do," Terry recalled. In addition to the physical attributes, Terry also remembered smelling a musky odor.

Terry sat in the boat for several seconds—though it seemed more like an eternity—as he watched the creature walk by. During this time it never looked back, presumably because it never heard the boy floating quietly on the pond. After the few eternal seconds, the creature finally walked over a bank and disappeared into a ravine leading to the creek bottom below. The creature was now completely out of sight, but Terry could still hear its footsteps in the leaves as it continued to walk. Shaken, Terry quickly paddled to the bank and got out of the boat. At that point, the creature's footsteps stopped. Terry stood frozen, listening intently in the direction of the ravine. Then he heard the creature start running!

"I didn't panic until I heard it start running," Terry confessed. From the tone of his voice, it's apparent the memory still chilled him. "It was pretty much nervousness until then, but once I heard it run, fear overwhelmed me and I took off running for the house."

Terry arrived at the house out of breath and still holding the boat paddle in his hand. He told his mother what he had seen, and she immediately called his father. Lloyd Sutton pulled into the driveway a short time later.

"Just as I parked, before I got out, Terry came out of the house and was standing near the edge of the patio with his hands over his face," Mr. Sutton told me. "I could tell immediately that there was something wrong. I had no idea what it was, but I could see that something was not right. I stepped out of my pickup and said, 'Terry what's the matter?' He said, 'Dad, I just saw the Fouke Monster!'"

After Terry calmed down and explained what he had seen, Mr. Sutton felt he should head down to the pond to investigate. He quickly gathered his .357 Magnum pistol, a 35mm camera, and two flashlights. "I asked Terry if he wanted to go with me, but I could tell he was a little hesitant. I told him it might be better to go on back down there and get it over with, or he would always want to shy away from there," Mr. Sutton explained. Terry was reluctant, but with his father's encouragement, he agreed. It didn't hurt that his father handed him the pistol either.

The Suttons walked the quarter mile down to the pond as dusk hovered heavy over Jonesville. They immediately looked for any signs of tracks, but found nothing in the dry blanket of leaves, nor on the water's edge. They followed the creature's path down into the ravine and checked along the creek bank. They found no tracks but did detect the faint remnants of a foul animal odor. As Mr. Sutton put it: "I can't describe it, nor have I ever smelled anything like it, as I can remember. It may have been a little agitated after hearing Terry bump the bank with his boat earlier and the running may have contributed to the odor, I don't know. But it was for real. You could walk either direction from that spot and it would go away. It was just lingering there in the moist air of the creek bottom."

By now, darkness had enveloped the woods, making it difficult to continue the investigation, but they walked back to the pond and looked around once more. While walking west along the bank, a large animal suddenly tore out the brush and ran in front of them just out of sight. Though they pointed their flashlights in the direction of the animal, they did not get a glimpse of it. So they decided to give chase. "I jumped across the water and took off after it with Terry right behind me," Mr. Sutton remembers. "We were making so much noise running, I would say: 'Stop!' We could hear it straight ahead of us and would go again. Each time we would stop, we could still hear it, but then the last time we stopped…

1981: The creature is seen near the Sutton family pond in Jonesville.

nothing. We walked around in circles from the last time we heard it, but we didn't hear or see anything again."

With that, the Suttons trekked through the darkness back to the house and spent the rest of the evening talking about the incident. Terry kept the incident a secret for a long time, but eventually mentioned it to some friends. He was ribbed, laughed at, and called a liar on many occasions, but he never changed his story. During my research for this book, several people told me about Terry's sighting, saying that his family was well respected and that I should try to follow up on it. Mr. Sutton was a longtime deacon in the local church, and although some people laughed off the incident, most felt that Terry and his father were not the type to make up stories, especially knowing how easily it could become a source of ridicule.

Photo of the actual pond where Terry Sutton saw the creature.
(Photo by Lloyd Sutton)

As I spoke with Terry about his sighting, I noted the tone in his voice as he told me of the events. This was not a man telling me an exciting story he had made up to entertain people; this was a man describing something that had happened to him, whether he liked it or not. The underlying fear that he must have felt that night as a young teenager alone in the woods still lingered as he spoke about the creature. I suppose nothing rules out that someone could have played hoax on him, but this would have to have been an extremely convincing suit to fool an experienced hunter sitting only 60 feet away. And who knew he was even there fishing that evening other than his parents? No, this does not seem like a hoax. This is a man telling me of what he actually saw that night so many years ago when he managed to get a rare glimpse of a shadowy creature.

Sutton's sighting certainly fits with the theory that the creature prefers to travel the local waterways of the Sulphur River Bottoms. Strange occurrences always seem to happen in proximity to the vast network of rivers and ponds in the area. In June of 1981, just five months after Terry's sighting, Jerry Wayne Scoggins claimed to have also seen the creature, this time along the Sulphur River south of Jonesville. As the story goes, he was going to check on his boat, which was tied up near the water. After re-securing it, he began to head back through the trees on his way home but stopped when he heard something splashing in the water. Figuring it must be a large alligator, he turned around to get a look, only to see a large, hairy man-like animal moving across the river. After splashing through the water, it walked up on the bank and disappeared into the trees.

Another near-water encounter took place in 1987. I heard the story from James "Peanut" Jones as we sat around his kitchen table one sweltering summer afternoon. Jones, who has lived on his family's property along the Sulphur River all of his life, was well aware of the monster legend, although he had never seen anything himself. And, as well, he had never been scared by anything in those swamps… until one moonless night in May. As he and his wife quietly rowed their boat along the edges of a pond searching for frogs to gig, they heard something large moving in the brush beyond. Jones shined a spotlight in the direction of the noise and was surprised to see a pair of brightly illuminated eyes staring back at him. Curious to know what kind of animal might be watching them, they rowed out into the pond to get a better view. After a few paddles, Jones could make out the shadowy shape of a large creature. As they got closer, the animal began to move in the trees just beyond the edge of the water. Suddenly, Jones and his wife became very frightened. The creature was upright, walking biped, as large or larger than a man. Inevitably, thoughts of the Fouke Monster popped into his mind, and that was enough to make him drop the light and begin rowing back toward their vehicles parked some distance away. Once they reached the bank, both Jones and his wife jumped out and pulled the boat ashore. They grabbed a few things and ran to his wife's four-wheeler, fired up the ignition, and headed up the trail out of the bottoms.

Did they overreact to something that could not be seen clearly? Jones has seen every kind of fowl, fish, and beast that inhabits the rugged swampland, and best he could tell, this was no ordinary denizen of the swamps. He doesn't claim it was the Fouke Monster, but it was scary enough to make him leave his own truck there until the following day. After seeing the strange animal, it was quicker to jump on his wife's four-wheeler and get out of the area than to drive his own truck back. Daylight would provide plenty of time to retrieve his truck. Plenty of time and safety.

BOGGY CREEK II

Ever since *The Legend of Boggy Creek* proved its worth as a moneymaker, film studios had been bugging Pierce to consider a sequel. "AIP had been after me for years to make one more *Boggy Creek*," Pierce revealed in the 1997 *Fangoria* interview. "But I really didn't want to do *Boggy Creek II*. I think it's probably my worst picture."

By 1984 Pierce had produced eight other films and his interest in the monster subject was not what it was in the early 1970s. He had capitalized on an instant in time at the outset of the monster's media heyday and that was that. As a director, he had moved on. But as we are all sometimes painfully aware, studio executives are more than willing to sully the name of a classic film in order to milk a few extra bucks out of the franchise. So Pierce finally relented and agreed to write and direct *Boggy Creek II*, which was released in 1985. Looking at the film today, there's not much evidence to refute Pierce's own criticism of the final product.

In this outing, Pierce casts himself in the lead role as Professor Brian C. "Doc" Lockart, an anthropologist who has an interest in tracking down the swamp-dwelling Sasquatch rumored to live in southern Arkansas. Lockart wrangles a group of three students—including his son, Chuck Pierce, who plays the part of the male student—and heads off toward the bayou to hunt for monsters. In route they stop off in Fouke to visit with some of the locals. One scene takes place in Fouke's convenience store where the locals warn Lockart against the dangers. The clerk is played by James Tennison, who played the landlord in the climactic Ford scene in the original Boggy Creek movie.

One of the most interesting scenes in *Boggy Creek II*—well, perhaps the *only* interesting scene—takes place as the group visits a local farm on the way out of town. In it, Pierce uses a flashback to describe a sighting of the Fouke Monster. This scene is the only one reminiscent of the docudrama style that made the original film work so well, and it was based on a real incident told to him prior to shooting the first film in 1971. As the story goes, the farmer took Pierce down to his barn where he carefully pointed out the location of his chilling encounter. As Pierce recalls:

> He told me, "When I came in here to milk, [the monster] was standin' there, looking at me. He stepped out, and the back barn door was open, and the light was at his back. He twisted his head and looked at me very curiously, backed up a couple of steps, and then started walkin' off across my back pasture toward the woods. The hair drooped off his shoulders, off his arms, and he didn't have any clothes on. The hair completely covered his body."

Even though the man was willing to share the story, he did not want anything to do with a movie. He did not want his name used;

he did not want it reenacted under a fictitious name. He wanted nothing to do with it. Perhaps this is why the scene did not appear in the original movie, although the unwillingness of the Fouke residents' to allow their names and stories to be used never stopped Pierce in other cases. Perhaps Pierce just never had time to recreate it. Either way, all these years later he finally found a place for it, giving *Boggy Creek II* its only flicker of authenticity.

The rest of the movie is comprised of a totally fictitious hunt for "Bigfoot," wherein Lockart and his young companions use primitive computer surveillance equipment to hone in on a swampy Sasquatch and eventually its offspring. The monster in this film lacks the mystery of the original, as it treads in full view. Pierce's marginal acting skills don't help matters either, which he readily admits: "I played too big a role in the picture." The two girls of the group bring way too much *Flashdance* to the forest, while his son Chuck chooses to spend most of his screen time shirtless, for some odd reason. Jimmy Clem, who plays in many Pierce films, appears as the dirty hermit named Crenshaw, but even his inspiring performance fails to provide any real spark even as the final scenes erupt in a fiery cabin blaze. These and other laughable elements provided perfect fodder for the award-winning comedy series, *Mystery Science Theater* 3000, which subjected the movie to its wise-cracking in episode 1006 of the series.

I had the good fortune of visiting Jimmy Clem at his home in Texarkana during my research for this book. I found him to be an extremely interesting and engaging person, who recalled both the good and bad aspects of the film. He did not dispute the fact that Pierce was not altogether happy about making the sequel, but he admitted they both enjoyed the movie-making process and the chance to work together again.

Regardless of the pay-off in fun or money, the film was pretty much a bust. And to confuse matters, the film's title, *Boggy Creek Part II*—or *The Barbaric Beast of Boggy Creek Part II*, as it is sometimes billed—seemed to forget the previous sequel, *Return to Boggy Creek*, although no one can blame them for trying to erase the memory of it from the moviegoers' consciousness. However, in the end, II might as well have been III. With its lackluster storyline, cheesy elements, and laughable creature, Pierce's effort wasn't much better than the previous sequel and failed to register much of a blip on the horror movie radar. Since the time when *The Legend of Boggy Creek* frightened audiences during the mid-seventies, horror fans had experienced *Halloween, The Shining, Alien, Evil Dead, An American Werewolf in London*, and many other great films of the genre. The bar had been raised considerably, and Pierce was not doing the Fouke Monster any favors by creating his own fake stories, even if he was portraying the beast as more dangerous and aggressive. The Fouke Monster was better off and certainly more intriguing for what he really was: a shadowy legend on the edge of civilization.

SHINE ON, YOU CRAZY MONSTER

It was 1990. It had been nearly 50 years since reports first began to suggest that a seven-foot-tall hairy creature might be roaming the woods near Jonesville, and two decades since something attacked the Fords in Fouke. During that time, interest in the phenomenon had waxed and waned, but like the moon it always returned, sometimes shining as bright as ever. If it hadn't been apparent before, it was certainly clear now… the Fouke Monster was not going away. Not ever.

While printed reports throughout the 1980s are hard to come by, the 1990s offered new, credible sightings that are somewhat easier to track down. This is not to say that there weren't as many reports during the 1980s, only that the newspapers had lost interest and the internet had yet to offer a new outlet for publishing. As a result, I would venture to guess that many word-of-mouth reports only circulated among the locals for a time. But by the 1990s, it seemed that a surge of Fouke Monster activity was taking place, much of which managed to find its way to the public one way or another.

The first sighting was recorded by the ever stalwart *Texarkana Gazette*. According to the report, two men from Oklahoma, Jim

Walls and Charles Humbert, were traveling north on Highway 71 on October 22, 1990. As they approached the Sulphur River Bridge around 8:30 a.m., they got wind of a pungent odor that was strong enough to make them pull over, thinking there might be a large dead animal in the vicinity. After stopping on the gravel shoulder, just before the bridge, something off to the right caught their eye. It was a tall, man-like creature covered in shaggy black hair and running across an open field toward the south riverbank, east of the highway. The men were shocked!

"He walked upright just like a human, not like a bear or gorilla," Jim Walls later told reporters. "I don't know what I saw, but I know it had to be made of flesh, blood, and bone."

The men estimated its height to be eight feet and its weight at approximately 400 pounds. They described its face as looking "much more human-like than a chimpanzee or a monkey."

After the alleged creature ran approximately 200 yards across the field, it paused for a moment and then leapt from the high riverbank towards the water. The men quickly drove across the bridge and circled back to see if they could get another look, but the creature was already gone. Walls assumed that it must have jumped into the river and disappeared below the surface, but this was pure speculation since they never actually saw it hit the water.

Amazed and excited, Walls and Humbert attempted to get permission from some of the local landowners so they could continue their search, but they were turned down in all cases. "We went driving down some dirt trails, but no one would let us on their property," Walls said. "They seemed more concerned about keeping outsiders away than about any monster."

The men were convinced that they had witnessed something truly strange and unidentified. "I don't know what it was, to tell you the truth," Humbert concluded. "But it chilled my bones."

The Walls and Humbert sighting was not unique. The article goes on to quote Sheriff H.L. Phillips, who said his office had received at least one sighting per month during 1990.

Two years later, one of the most remarkable eyewitnesses sightings of the Fouke Monster occurred. In this case, the creature was not seen by just one, or even two people, but six people at once.

The incident took place on a cold, foggy night in October of 1992. At around 11:00 p.m. five young men were driving near Fouke on a lonely stretch of FM 134. Just after they passed County Road 8, near the McKinney Bayou, they noticed the bright headlights of a semi-truck coming toward them. As the truck got closer, they suddenly caught a glimpse of someone, or some*thing*, as it came out of the shadows and walked across the road in full view of both vehicles. The figure formed a dark silhouette in the headlights, but they could still see it was some kind of large hairy animal walking upright on two legs. The creature had come from the thick woods that lined one side of the road and was headed toward an open field that lay on the opposite side. Startled, both drivers hit the brakes.

Their first impression was that it must surely be a bear, but as they continued to watch the creature stroll across the road a mere 50 yards away, they could tell it was not. It was more man-like, but larger than a man, standing an estimated seven feet tall and walking on two legs the entire time. It never paused or looked at the men; it just kept moving. Facial features and other smaller details were hard to see because of the silhouette effect, but they were certain the creature was not an ordinary animal.

"It was definitely taller and thicker than a man, and bushy like it had a thick coat of hair," Rusty Anderson told me during an interview. He was one of the five men in the car that night, two of whom I was fortunate to speak with. "It was one of those things that made the hair stand up on the back of your neck," he added, as he recapped the events of that night.

After taking a few long strides, the figure left the road and entered a field where it moved beyond the reach of the headlights. At that point, the truck driver hurriedly got out of his vehicle, as did the five young men. They tried to get another glimpse of the strange creature, but it had already disappeared into the night.

The men spoke with the trucker, all of whom were astounded by what they had just seen. "I remember that I felt in awe," the other witness told me. (He was driving the car that night, but

1992: A strange creature is seen by six witnesses near McKinney Bayou just south of Fouke.

wishes to remain anonymous.) He and the other four passengers, including Rusty Anderson, lived in the general area, so they had certainly heard stories of the Fouke Monster many times and felt that perhaps they had just witnessed the animal for themselves. The truck driver, however, had never heard the legend. He was absolutely amazed by the back-story.

The incident was so surreal they discussed the possibility of having been hoaxed. But in the end, the witnesses did not feel this was the case. They were certain it was a real animal. "The reason I believe it's true, and not just someone pulling a stunt, is the fact that there are no houses out there," the witness stated. "The field it went into is at least four or five miles square. We were in the middle of nowhere, and there are few cars that travel that road."

After a last look into the field, the parties returned to their vehicles and went their separate ways. The driver of the car, who eventually discussed the sighting with a friend of mine in 2004, was leery about telling anyone at first.

"All of us talked about what we saw to our families. I still think they don't believe us. We never reported anything at first, because we didn't want people thinking that we were trying to get attention or something."

Considering the number of witnesses in this case, it seems unlikely that they had all been hallucinating. The two witnesses I spoke to seemed credible and I did not feel they were exaggerating or making up the story. Of course that doesn't rule out trickery, but if they had been fooled by a clever hoax, then the hoaxer was indeed a brave person. Anyone who lives near Fouke knows full well that the area is stocked with hunters and woodsmen. The possibility of getting shot makes hoaxing a *very* risky affair.

When asked if they thought to go back later and look for evidence, such as tracks, the witnesses admitted that unfortunately that did not occur to them at the time. As such, the sighting remains a true mystery, yet one that stands out because of the number of people involved in the case.

But it was not the only time the creature found itself caught in headlights. According to Dorothy Briggs, who worked at the local convenience store in Fouke, two gentlemen came in one night well after midnight, sometime in the early 1990s. They had been traveling on Highway 71 in route to Shreveport, when they spotted something very strange crossing the road near Fouke. The two men, one of whom was Roosevelt Shine from Memphis, claimed to have no prior knowledge of the Fouke Monster, which made the sighting that much more incredible.

As the two men drove south along Highway 71, they saw what they described as "some kind of ape-man" walk out of the woods and onto the road in front of them. Mr. Shine slowed the vehicle until they rolled to stop about one hundred feet from the creature, which had paused on the shoulder of the road, eyeing the approaching headlights. As they sat there in growing disbelief, they could see that the creature was definitely walking on two legs, covered with dark hair, and standing approximately seven feet tall. It seemed adept in its movements, so it was fairly clear that it was some sort of bipedal animal and not merely a bear that had taken to walking upright.

Moments later, another car approached from the opposite direction and came to a stop as the occupants also noticed the strange creature frozen like a deer in the headlights. The couple in the car were from Fouke and well aware of its namesake monster, but they never believed they would see such a thing. The creature continued to stand at the side of the road, wary of the growing audience, when incredibly, a semi-truck came up alongside the Fouke couple and stopped to see what was happening. The witnesses now totaled five adults.

A few moments later, the creature darted back towards the woods, running fast on two legs. Immediately, all five people got out of their parked vehicles and began to discuss what had just taken place. To everyone's amazement, they all saw what was very clearly a creature fitting the description of the Fouke Monster. After some conversation, they decided to report the incident to the local authorities, which they did. Afterwards, Mr. Shine and his companion made their way to the convenience store in Fouke where they told their story to the clerk, who eventually put them in touch with Smokey Crabtree.

So here again we have a case in which the creature was seen in plain view by multiple witnesses. Just as in the October 1992 case, something comprised of flesh and blood, walking on two legs, came out of the woods and approached the road. All these people are not likely to mistake a bear, deer, panther, or any other animal for a bipedal ape-like entity. So if misidentification is ruled out, then it only leaves two options: either it was a hoax perpetrated by some individual who did not care about the risk of getting shot or run over... or it was a bona fide mystery animal.

6. BONES AND SHADOWS

A GLIMPSE BEHIND THE CURTAIN

I haven't been lucky enough to witness an ape-like creature cross the road, but I've seen some strange things in my life. I spent two decades touring as a rock musician, playing all across the United States, Canada, and numerous European countries. I've been in nightclubs made out of old World War II bunkers, visited dungeons in England, strolled through spooky graveyards in Austria, and performed in the hallowed rock halls of CBGB's club in New York. I've mingled with all kinds of unique people along the way... musicians, movie stars, monster hunters, fire breathers, sword swallowers, contortionists and even a human lizard man. I've visited the International Cryptozoology Museum in Portland, Maine; been to the Museum of the Weird in Austin, Texas; and come across a good many Bigfoot researchers who have shown me some very curious items that defy simple explanation. But nothing compares to the time I was introduced to a huge decaying skeleton locked away in an old building on the outskirts of Fouke, Arkansas.

It was a warm day in April. My wife and I had made our way to Texarkana, where we were guests at a meeting of Bigfoot researchers being held at a local outdoors shop. I was excited because Smokey Crabtree was scheduled to speak. I had met Smokey six months earlier at a related event, where I purchased one of his books and spoke to him briefly, but I had never seen him speak about the Fouke Monster.

Smokey spoke for nearly an hour, carefully going through the details of his son Lynn's encounter and entertaining the group with stories about growing up in Mercer Bayou. After the meeting, we were invited to accompany Smokey and some others to an undisclosed location where we could see some interesting relics having to do with the alleged monster. Naturally, I jumped at the opportunity.

Following by car, we eventually arrived at the designated location and parked. After the group gathered around, Smokey led us to a nearby building where he fiddled with a lock and latch until finally its heavy door slid open. We followed him into the dimly lit interior as he led us toward the far end, navigating through a haphazard collection of what appeared to be old tools, discarded furniture, and other items being stored there. With our eyes still adjusting to the low light after coming out of full sun, we practically had to hold on to each other's shirttails to avoid tripping. It was like a group moving through one of those haunted house attractions at Halloween.

Eventually we came to a large, free-standing wooden panel that was located near the back wall. The panel bore a crudely painted message that read: "8-foot skeleton." It looked like something you might see at an old traveling sideshow.

By now I had my suspicions as to what we were about to see and I could feel a twinge of true excitement sparking in my brain. The whole scenario reminded me of a scene from the old 1972 made-for-television movie, *Gargoyles*. In the movie, a writer and his daughter are summoned to a dusty roadside attraction called Uncle Willie's Desert Museum to investigate a strange artifact, which the owner claims to have found in the surrounding hills. At first, old Uncle Willie seems like a charlatan, but eventually he takes the pair out to a shed where he's kept the artifact. After making them promise not to take photos, he throws back a black curtain to reveal a bizarre looking skeleton that is roughly human size, yet possesses a horned skull, sharp beak, and large wings. That cinematic moment has stuck with me since I first saw the movie as a kid, so it seemed extremely surreal that years later I would find myself living a very similar scene. To further the irony, *Gargoyles* was released in 1972, the same year as *The Legend of Boggy Creek*. The parallels were rather eerie.

Smokey slid the wooden panel out of the way, revealing a large display case behind it. The case appeared to be roughly eight feet wide, four feet deep, and two feet high. It was made of clear acrylic glass resting on top of a wooden support structure, which was like a table on wheels. Smokey undid the front latch and propped open the glass lid so we could get a better look inside.

The first thing that hit me was the smell—the stench of organic decay made worse by the closed confines of the acrylic glass case and dead air of the musty building. When the lid was raised, all the pent up odor crawled out and hung like an invisible fog in the immediate area. It was pretty much the worst thing I have ever smelled.

But as I gazed into the case, I quickly forgot about the horrible smell. Inside lay the skeletal remains of something that had once been a very large animal… of some type. It was devoid of skin but covered with a layer of dried sinew, tendons, and muscle which still clung to the bones. It was something between a skinless corpse and a skeleton, mostly brown and yellow in color and completely intact except for the skull.

Everyone in the group stood somewhat aghast, adjusting to the foul odor and trying to figure out what kind of skeleton it was that lay before us. I ran through a quick list in my mind. Human… no; dog… no; deer… no; horse… no; bear… I don't think so; ape… maybe. Smokey explained that it had been found just south of the Arkansas border in Texas. I amended my list of possibilities to include more of the indigenous wildlife such as armadillo, beaver, and coyote, but the skeleton was far too large to be any of those. Cougar, perhaps? Yes, cougars had been known to inhabit the area, but still it looked too massive even for the largest variety of Southern mountain lion. I was at a loss for a reasonable explanation. All that was left at the moment was to consider the possibility that it had belonged to something more unique… something as yet *unexplained*. Was I looking at the skeletal remains of a Fouke Monster? My nerves tingled at the thought.

Naturally the group began to ask questions and Smokey finally gave us a brief rundown of the back-story. The skeleton had come into his possession nearly 20 years earlier, back in 1991. Two men had discovered the carcass near the border of Texas and Louisiana. It was headless and skinless, but otherwise in good shape for remains. The men were familiar with the stories about a mysterious creature that stalked the woods near Boggy Creek and remembered the name Smokey Crabtree. Thinking that the skeleton could perhaps be that of some unknown animal fitting the general description of the Fouke Monster, they collected the corpse and immediately contacted Smokey.

Intrigued by the news, Smokey made his way to Vivian, Louisiana, where the skeleton had been relocated. After looking it over, he agreed that it resembled the Fouke Monster and thought it would be a good idea to preserve and subject it to further testing. Smokey entered into an agreement with the men that gave him a full one-third share of any future profits from the skeleton, as well as the right to show it publicly. Smokey went on to tell us that the remains had been examined by several university professors and scientists, and that some basic DNA tests had been run to the tune of nearly $10,000 dollars. The only definite conclusion they could make was that "it wasn't human." He had theories as to what kind of animal it might be, but he stopped short of going into further detail.

I turned my eyes back to the Plexiglas coffin before me. The headless animal lay on its side with its legs extended and its knees slightly bent. Its arms were sprawled out in front, bent slightly at what would be considered the elbow. Its back was to us, as it faced — so to speak — the rear wall of the case. I looked closely at the hands and feet. They looked elongated and primate-like, at least to my untrained eye. There were four fingers on each of the hands and four toes on each foot, with the feet looking to be at least 14-16 inches in length. I tried to estimate the creature's height if it were standing beside me. It would be roughly seven or eight feet tall... the same as the reported height of Fouke's mystery creature. I looked closely at the shoulder blades, which folded inward with the narrow frame of its ribcage. Given that anatomy, it would conceivably have walked in a hunched fashion... if it were to walk bipedally. Again, just as some of the sighting reports had described.

The desire to take a photo burned within me. My cellphone was handy, but I have too much respect for Smokey and therefore honored his wishes, despite the paparazzi urge. Smokey explained that one reason he couldn't allow photos was because of the legal agreement he had with other interested parties. The skeleton itself was strange, and it was evident that the circumstances surrounding it may be just as odd.

I looked at the carcass one last time and turned to join the rest of the group as they made their way back through the trail of "antiques" toward the exit. The doorway glowed on the far side of the building as sunlight spilled inward with copious amount of airborne dust. As I stepped out of the darkness and returned to the warm sun, I felt as though I had seen something rare and special. Something that was hidden from the world, tucked away in a small corner of obscurity on the edge of commonly accepted science. In a way I felt honored to have seen it. As if I had been given a rare glimpse behind a magic curtain. It may sound melodramatic, but the whole experience was very surreal at the time, given my lifelong interest in the Fouke Monster and general love of all things creepy and mysterious. I might have thought it to be a mere dream, if not for the foul smell that still lingered in my nose.

STRANGE REMAINS

At the time I saw Smokey's skeleton, I had not yet been overcome with the idea of writing a book on the Fouke Monster. At that point I was looking into the local "Bigfoot scene" out of both personal curiosity and as research for some articles I was writing for the popular horror magazine, *Rue Morgue*. By a strange set of circumstances, which I jokingly refer to as *destiny*, I had scored a freelance position with *Rue Morgue* to cover cryptozoology-related horror movies, so naturally I was eager to brush up on my Fouke-lore since rumors of a *Boggy Creek* remake were circulating at the time. If there was anybody qualified to cover the new movie, and to perhaps write an accompanying retrospective on the original *Legend of Boggy Creek*, I felt that would be me. So I considered myself fortunate to have been able to visit with Smokey and to have seen his skeleton, of which I had only heard rumors about prior to that day.

After a bit of research, it didn't take long to find out that the skeleton had long been a source of mystery, debate, and misinformation among Bigfoot researchers worldwide. Threads about the subject could be found on various cryptozoology internet forums and a handful of other sites having to do with general Arkansas folklore or the Fouke Monster itself. The postings ranged from unanswered questions to incomplete theories, to outright wild guesses and incorrect details. The exact composition of the specimen had never been determined, or at least divulged, so to the average cryptid enthusiast this was an important lingering question. And rightly so. The possibility that somewhere in the winding woods near Fouke, the skeletal remains of an undocumented ape-like creature might be locked away was promising. We've all seen numerous grainy video clips of alleged Sasquatch, while researchers have turned up various strands of suspect hair, countless plaster casts of giant feet, and the occasional sound byte of a howl, but rarely had an actual bone been brought forward—much less a nearly complete skeleton—that, even for a scant moment held a real grain of hope that it was from such a creature. So with the monster-filled reputation of Fouke behind it and the association of the Crabtree name, it was only natural that rumors of such a specimen would spark widespread interest.

But as I continued to dig for answers and ask around, I started to wonder why there was still so much debate and misinformation over the issue. Not only does Smokey cover the topic in one of his books, but amazingly, the skeleton is the centerpiece of a low-budget documentary called *The Hunt For Bigfoot*, produced in 1995 by none other than Jim McCullough Sr. of *Creature From Black Lake* fame. In addition, I have been able to track down old newspaper articles that provide convincing details as to the true origin of the mystery skeleton. Granted, all of these are hard to come by—the book is mostly available through Smokey himself, the documentary is extremely rare and hard-to-find, and the news articles have been all but lost until now. It's also why I found myself standing in a room with the skeleton without much foreknowledge of its existence.

The story of the skeleton actually begins in November of 1991. According to a news article that ran in the *Marshall News Messenger* out of Marshall, Texas, the remains were initially discovered on private property near Karnack, Texas, by James Mackey and a friend. Mackey, the owner of said property, had stumbled across the shocking sight on November 27. According to the report, Mackey quickly notified the Harrison County Sheriff's Office, which sent Captain Frank Garrett to investigate. Upon seeing the bones, Garrett arranged for them to be picked up within 24 hours and examined. But in the meantime, Mackey's 17-year-old nephew decided it would be a better idea to transport the bones to Vivian, Louisiana, for safekeeping just in case they were from some undocumented animal, such as the one that had been seen around Fouke. The teenager, apparently aware of the legend of Boggy Creek and Smokey's association with it, contacted Smokey and told him what they had found. Smokey quickly made his way to Vivian in order to evaluate the discovery.

Once there, it is evident in both Smokey's account and the newspaper accounts that he was impressed enough with the gruesome specimen to pursue partial ownership of it. Though its feet were not likely to have made the so-called three-toed tracks that had become associated with the Fouke Monster, the skeleton had enough similarities with it to pursue the question. According to an article printed on December 11 in the *Texarkana Gazette*, Smokey had this to say: "I told them that it looked like it had some historical value since it resembled what's been seen around here. It had enough human features that I felt it needed to be preserved."

Smokey didn't jump to conclusions but made no bones about the fact that he wished to keep it. As he states in his memoir, *Too Close to the Mirror*, "It was evident to me that the skeleton was the nearest thing to being the remains of the Fouke Monster that I had ever seen." As a result, he proposed a three-way deal of ownership with the teenager and another unnamed man. The three men signed an official contract saying they would split the money in case it did turn out to be something of unexplained origin and somehow turned a profit.

The skeleton was then transported to Fouke where Smokey began a process of trying to preserve it as best he could.[13] One of the steps was to build the glass enclosure, the very one I had seen. Another was to have it examined by a scientist from the University of Arkansas in Little Rock in an attempt to identify the animal involved. Unfortunately, nothing came of his initial visit despite the fact that the scientist took samples for proposed DNA testing.

Since the late 1980s, Smokey had been running a local event called the Monster City Jamboree. The jamboree was a music show that took place every Saturday night in a converted grocery store that Smokey owned on the north end of Fouke. The show featured a house band that played country tunes for a crowd sometimes numbering as many as 500. By the early 1990s, the jamboree had become so successful that it was, as Smokey put it, "the number one show in the four states area." Because of this built-in audience, Smokey decided it would be a great way to raise money for the scientific testing of the skeleton. He had sealed the glass case, so as to minimize the foul odor and help with preservation. That also made it reasonably fit for public appearances.

After some additional preparation, he began to show the remains at the jamboree to anyone willing to fork over three dollars. A good many music fans stayed late to see the moldy old skeleton, but for the most part it only caused problems because of the morbid subject matter. As well, the news media descended on it like human flies clamoring to get the skinny on the animal, be it a monster or otherwise. News stories began to air on both Texarkana and Shreveport television stations, although bereft of any photos since they did not have permission from Smokey to show an image of the skeleton. In this case, the subject matter was left up to the imagination of the viewer.

The public exposure led to other oddball inquiries, one in particular coming from a person claiming to be a representative from the Smithsonian Institute. Two men also showed up claiming to be the original owners, but Smokey sent them all away empty

*1991: Strange remains found in the woods approximately
44 miles south of Fouke.*

handed. The scene was reminiscent of the old days when monster hunters had, not once, but twice wrecked havoc on Smokey's personal life and property as they tried to track the monster. Such is the price of being a local celebrity, one might say, but that didn't make it any easier to deal with the backlash.

Reluctantly, Smokey was forced to return the skeleton to its quiet slumber behind locked doors. He was still no closer to a final determination as to its identity, although rumors about its true nature had been flying since the beginning. It was only a matter of time before the cat, so to speak, would be out of the bag.

SKELETON IN THE CLOSET

A month or so after I first saw the skeleton, I purchased Smokey's second book, *Too Close to the Mirror*, which features a clear photo of the skeleton's feet on the back cover. The photo is in full color, depicting the lower portion of the skeleton's legs from about the shins down. The bones are laying on a bed of grass and dirt, so it was evident that the photo must have been taken at the time the carcass was discovered.

After viewing the actual remains, one of our group commented that they thought the bones might belong to some kind of cat. With the photo to compare, I did a quick internet search for any images of mountain lion (i.e., cougar) skeletons, as these kind of large cats would be the type found in Texas. The results were thin, but there was at least one fairly good image of a mountain lion skeleton that I could compare to the feet on the back of Smokey's book. I saved the image to my computer and rotated it a bit to match the orientation of the feet from Smokey's photo. My heart sank a little as I stared at the two images. There were indeed many similarities that could not be ignored, although this was by no means a solid scientific conclusion.

After finally reading Smokey's book, I came to realize that it was no secret that the remains were thought to be those of a large cat, specifically a Siberian tiger. Smokey mentions the particular rumor,

pointing out the paradox of its growing value. In his own words: "...
I had never been thoroughly convinced as to what I had in my
possession. Experts were wanting to tell me it was the skeleton of a
Siberian tiger, others wanted to buy the skeleton, offering as much
as sixty five hundred dollars. If the skeleton was that of a Siberian
tiger why were they willing to offer such an amount for the bones?"

Perhaps the skeleton's perceived worth had gone beyond the
question of whether it was actually a cat or not. Within a short time
the specimen had already woven itself into Fouke Monster legend,
becoming valued more for its part in history than for its biology. In
other words, it became a part of Fouke Monster history in spite of
itself. But like Smokey, I maintained hope... that is until I managed
to track down two items that shed a bright light on the whole
mystery. One of these was a nearly three minute long exposé, which
had aired in 1991 on Shreveport's KTBS television station. The
other was an article that ran in the *Texarkana Gazette* around the
time of the skeleton's discovery. These reports provided the missing
pieces needed to recreate the historical chronology of the skeleton
saga, and also indicated, much to my disappointment, that it was
most likely *not* the remains of an undocumented bipedal primate,
which may or may not be indigenous to the swamps of the southern
United States. In short, I was happy to have the whole intriguing
story but disappointed by its less-than-exciting conclusion.

The newspaper article ran on December 11, 1991, with the
headline: "Boggy Creek remains apparently Siberian tiger." The
report cites the original story published by the *Marshall News
Messenger* and goes on to add new details regarding the skeleton's
appearance at the Monster City Jamboree. The article lays out the
twisted tale:

> ... Capt. Frank Garrett, Criminal Investigations
> Division supervisor for the Harrison County Sheriff's
> Office, said the remains are those of a Siberian tiger.
> Garrett confirmed information in a story published
> by the *Marshall News Messenger* in Marshall, Texas, that
> the remains originated in Jefferson, Texas, and eventually
> made their way to Fouke.

> The tiger, which was reported to have died of pneumonia, belonged to a Jefferson resident who raises exotic animals. The exotic animal owner was reported to have sold the dead animal to another resident who, in turn, sold it to a local taxidermist.
>
> The taxidermist beheaded, declawed and skinned the tiger. Tiger bones and other remains were discarded by the taxidermist on nearby private property belonging to James Mackey of Shreveport, La., Garrett said.
>
> Mackey said he and a friend discovered the bones on his property near Karnack, Texas, Nov. 27. Mackey said he notified the sheriff's office, which arranged to examine the bones on Nov. 29.
>
> But Mackey said his 17-year-old nephew had, in the meantime, transported the bones to Vivian, La., near the Arkansas border—apparently aware the Boggy Creek Monster legend originated Fouke.
>
> The nephew eventually met Smokie Crabtree [sic]. From there, the remains apparently made their way to Fouke.

The trail of ownership sounds like something out of a gruesome sit-com, but judging from the report, it seems that the Fouke Monster may have a boney imposter. To verify the story, I contacted Captain Frank Garrett, now retired from the Harrison County Sheriff's Office. He recalled the incident immediately, remarking that: "it's one of those stories that sticks with you."

Garrett was well aware of the controversy surrounding the skeleton but assured me that the details offered in the news accounts were accurate, and that the bones had a traceable history to their origins as a living Siberian tiger. According to Garrett, the cat had been living at one of the local exotic animal reserves in Jefferson, Texas, until it died from complications brought on by pneumonia. The remains were sold to an individual in Marshall, Texas, by the name of Wayne Scoggins,[14] who took them to a taxidermist for mounting. The taxidermist removed the head, skin, tail, and claws and dumped the rest of the remains on the property of James Mackey where they were discovered by two hunters. Garrett and a local veterinarian examined the bones, and satisfied that they were not human, arranged for them to be picked up the following morning. However, before they could be removed by officials, the bones were taken by Mackey's nephew and transported to Vivian.

"It looked like an upright creature because it had no paws on it," Garrett remarked, understanding how it could be mistaken for something other than what it was. "What [the taxidermist] had done was take the claws and everything off of it. You had part of a foot, but you didn't have all of it. And that was the same thing with the front and the rear [legs]."

Granted, no DNA test has offered conclusive proof of the skeleton's true biological identity, even to this day, so the possibility for error is still there, but the logical conclusion points to something more mundane. The remains are indeed exotic and would qualify as a cryptid if the animal had been roaming the Texas woods alive, but unfortunately they are most likely not something so exotic as to be those of an undocumented ape-like creature.

Regardless, the skeleton seems to possess an undying appeal to those who have an affinity for the Fouke Monster. And soon enough, its creepy charisma would call again from that cold, musty coffin.

THE HUNT FOR BIGFOOT

Ever since its debut at the Monster City Jamboree, word about Fouke's mystery skeleton had been buzzing around the cryptozoology wires. Among the listeners were Jim McCullough Senior and Junior, better known as the creators of *Creature From Black Lake* some 20 years earlier. Not wanting to miss a good opportunity to capitalize on Fouke Monster fame, the McCulloughs hatched a plan to exploit the skeleton. The idea was to film a low-budget documentary about Bigfoot and feature the mystery skeleton as its centerpiece.

To accomplish this would of course require the cooperation of Smokey, so the McCulloughs promptly called him up with a proposition. According to Smokey's second memoir, *Too Close to the Mirror*, he agreed to a tidy sum of $5,000 to let the skeleton be photographed and tested. In addition to the filming rights, the contract gave the McCulloughs one-forth interest in the skeleton. Smokey was also told that the movie would be titled "The Hunt For Bigfoot with Smokey Crabtree," in order to further tie the skeleton's owner to the production. Up until that point, no one had been given permission to take so much as single photo of the rotting remains, so naturally the specimen had built a buzz in the Bigfoot community. The McCulloughs hoped this would translate into sales of the final product.

After working out the details, the shooting commenced. Both McCullough Senior and Junior worked on the project, filming much of it in and around Smokey's home near Days Creek. The final product, released in 1995 with the shortened title *The Hunt For Bigfoot*, is ultimately a haphazard pseudo-documentary/movie that fails to provide much gratification, regarding either the general subject matter or the skeleton itself. Presumably because of its low budget production, the documentary never received wide market circulation, making it a rare commodity even for cryptozoology enthusiasts.

The documentary is narrated by host, Clu Gulager, best known as a prolific television actor who appeared on hundreds of shows including *Kung Fu, Ironside, Hawaii Five-O*, and *Knight Rider*, along with a few notable horror movies such as the classic *Return of The Living Dead*. He's a personable actor with a distinctive voice, but unfortunately, in this case, the badly lit set and African safari get-up only downplay his talent.

The documentary comments on the Bigfoot phenomenon in general but concentrates on sightings and stories from the Arkansas/Texas area. These accounts are intercut with statements and opinions from various people, including an alleged psychic and a doctor. The psychic, Valorie Taylor, ventures to guess that Bigfoot creatures live in caves in sort of a "Cro-Magnon man kind of situation." Dr. David Otto also provides commentary, but it centers mostly on man-ape legends rather than anything hard science.

VHS packaging for "The Hunt For Bigfoot" Documentary.

Several residents of Fouke are interviewed throughout, one of them a woman who works at an unnamed convenience store. On screen, she recalls a time several years back when two men came into the store saying they had seen "the creature" (i.e., the Fouke Monster) at about 3:00 a.m. As she put it, they were "shaking, white as a ghost." The store is obviously the Monster Mart and the men were Roosevelt Shine and his friend, whose account I have already covered. Unfortunately, the details of their account are not discussed in the documentary, which is a shame since it would have made for compelling content.

The documentary goes on to highlight *Creature From Black Lake*, using footage from the movie to round out the production. However, since that movie is fictional, it adds nothing to the "scientific hunt for Bigfoot" and merely comes off as cheap filler material. To make it even more cheesy, a "Fouke Monster hunt" is staged by filming an actual family being guided into the Mercer Bayou by Smokey himself. The family was visiting from Michigan and had no scientific background or prior experience hunting for Fouke's undocumented animal. They mostly walk around looking confused, as Smokey scouts ahead on the river in his pirogue. The whole thing is rather silly and a discredit to Smokey's serious, life-long pursuit of the creature.

Throughout *The Hunt For Bigfoot* the narrator drops hints about the mysterious skeleton, implying that something big might be revealed when scientists examine it for the first time. It is only towards the end of the documentary that the narrator finally focuses on this piece of alleged evidence. A short back-history is provided, telling the viewer about how it was found in Texas and ended up in the hands of Smokey Crabtree. The skeleton is first examined by Dr. Beth Leuke and Dr. Vaughn Langman, both professors of biology. They take various measurements and make conjectures based on its current condition and the state in which it was reportedly found. Leuke makes a good case against scavengers having taken only the head, which, ironically, backs up the official police report that it was dumped by a taxidermist.

Following this evaluation, the skeleton is transported to Bossier City, Louisiana, where it is examined by forensic anthropologist Clay Stewart. However, as in the case of the previous two examiners, nothing conclusive is offered.

In the end, it seems that the original report filed by the Harrison County Sheriff's Office — that it was the carcass of an exotic Siberian tiger pet — was still the most likely explanation. The documentary merely exploits the interesting looking remains by leaving out the police report details and not bringing in an expert more familiar with the biology of wild cats.

The production apparently floundered for a while as the McCulloughs sought distribution. It was eventually released in 1995, but by this time the McCulloughs and Crabtree were in a dispute over money issues. As with Pierce and *The Legend of Boggy Creek*, Smokey felt he needed to take legal action and found himself right back in the offices of a law firm to sort it out. Twenty years had passed since the Pierce episode, but it was evident that filmmakers and Fouke just didn't seem to mix very well.

MONSTERS BY MOONLIGHT

With the coming of the world wide web, reports of Fouke Monster sightings were easier to propagate. The stories no longer needed a traditional printed outlet, such as newspapers or books, to reach the public. Reports could be investigated by Sasquatch researchers, bloggers, or cryptozoologists, and posted for the world to see with the click of a few keystrokes. Using this technology, innovative special interest groups began setting up websites and casting their URL's into the vast cyber sea. One such group was the Gulf Coast Bigfoot Research Organization, GCBRO for short, founded in 1997 by Bobby Hamilton.

Most people associate Bigfoot with sightings in the Pacific Northwest of the United States or in British Columbia, Canada. Only *Fate* magazine or the local papers would pay much attention to any of the local varieties, unless of course they managed to star in a movie. But even for the Fouke Monster and his Southern kin, there was no official place dedicated to furthering alleged reports of the Southern Sasquatch until organizations like the GCBRO changed the situation. As stated on their website, one of the group's primary missions is "to bring to light more encounters and happenings from the South."

Over the years, several notable researchers have joined the GCBRO, including Chester Moore, Jr., who has penned several cryptozoology-related books of his own and is a frequent contributor to magazines such as *Texas Parks & Wildlife* and *Louisiana Sportsman*. The GCBRO database holds numerous sighting reports, many of which have been investigated by one or more members of the team in an effort to maintain a reasonable level of integrity. Scanning the entries, it is not surprising that a couple of the more interesting reports originated from the Sulphur River area near Fouke.

One of the incidents occurred on September 17, 1997, as reported by a woman living 10 miles south of Texarkana on Highway 237. Her property, which she and her husband purchased only weeks before, was north of Boggy Creek on the west side of Fouke. She stated that while she was out one evening shopping, her husband was outside tending to the lawn and working on his car. At around 6:00 p.m., he began to feel uneasy. It was as if he were being watched, but initially he could not figure out the source of the strange feeling. It wasn't until he began look into the surrounding wooded area that he noticed something moving in the brush. When he walked closer to the woods, he saw a "large, dark brown creature." He described it as "about seven feet tall with a dark black face." Unsure of what it was, the man stood there and watched the creature for several minutes, until he decided to return to the house and grab his pistol. But when he returned, the creature was no longer visible, so he walked slowly toward the woods in an attempt to get another look. Finally, he spotted it sitting on the ground about 30 feet from where he had first seen it. By his estimation, he was roughly 200 feet from it. He stood there for sometime, watching the animal, until finally the sun dropped behind the horizon and the animal was engulfed by the encroaching darkness. When the woman returned from shopping, he was waiting in the yard with his gun and a flashlight to make sure she got into the house safely.

This was not the only strange incident to occur near the couple's home, however.

On the bright moonlit night of February 2, 1999, the couple was sitting outside, watching for the heard of wild hogs that often passed through their 50 acre property in hopes of bagging a tasty one for cooking. To get a better shot, they moved up to the second floor of their two-story home. As the hogs approached, they could see their wily silhouettes against the moonlit backdrop.

Suddenly, they heard "a strange sound from far to the right." Following that came another sound, seemingly in reply to the first, from directly behind the hogs. "It sounded like a whistle that turned into a gibberish type noise," the woman reported. Frightened, the hogs snorted and scrambled for the treeline. "Then there was a blood curdling scream," she continued. "I could tell by the sound that one of the hogs was being slaughtered. It was followed by loud shriekish screams and howls with this gibberish sound mixed in. I have never heard anything quite like this before. It made the hair stand up on the back of my neck."

The noises continued for another five minutes, inspiring all the neighboring dogs to go crazy. Then there was silence.

The next morning, the couple found a large, flat spot in the field, approximately four feet in diameter, covered in fresh blood. The grass in the surrounding area was torn up, indicating it had been the scene of at least one hog's final stand, although no dead hogs could be found. "Whatever killed that hog, picked it up and walked away with it," the wife concluded.

While no hairy creatures were observed in this case, images of the Fouke Monster spring to mind as both the Fouke Monster, and Sasquatch in general, have been accused of preying on hogs. This case, if true, seems to implicate two or more large creatures working together in a calculated game of hunt with the hogs as prey. But nothing about that night is certain except for the hog butchery, so this is purely conjecture. The area is known to host mountain lions and other large predators, so the identity of the actual killer will never be known.

Another report posted in the GCBRO database occurred around the same time, although the creature was spotted in the morning sun instead of by moonlight. It was July 11, 1998, around 9:00 a.m. A woman (who requested anonymity) was babysitting her sister's children in the Jonesville area southwest of Fouke. Eager to enjoy the daylight before the unbearable heat set in, she walked the children a short distance from the house to an area where some new timber had been cut, when something strange caught her eye off to the right. As she focused on it, much to her horror she realized it was a "very large hairy creature" watching them from the edge of the woods. The woman was shaken but managed to keep her composure as she hurried the kids back toward the house. "I kept looking out the corner of my eye to see if it tried to follow us," she wrote, as she described the sense of fear. "I have to say this, that was the longest walk I have ever had to take. I was scared to death and was so afraid that the thing was going to come after us. But it never did."

The incident bears a striking similarity to one of the memorable scenes in *The Legend of Boggy Creek* where some kids and their mother witness the beast as it stands on the edge of a clearing. If true, it would seem that the monster was still up to his old tricks, casting a curious eye upon the human presence near his old haunting grounds of Jonesville. In 1998 the area was still much the same as it had been back in the 1960s.

Another report from the area was investigated by my friend Conor Ameigh of the Bigfoot Field Researchers Organization (BFRO). The incident occurred in the late night hours of November 1996. In this case a truck driver was traveling north on Highway 71, having made an earlier delivery in Shreveport. The sky was overcast, but the Halogen headlights of his 18-wheeler effectively carved away the darkness as he rolled across the lonely blacktop. The trip had been uneventful until approximately 1:00 a.m., when he neared the Sulphur River Bridge that crosses the highway eight miles south of Fouke. As he drew closer, his eyes were drawn to a tall dark shape standing near the water on the right-hand side of the road. At first he believed it to be a dead tree trunk, but then he noticed the shape had two red eyes reflecting back in the headlights. Suddenly the "tree" began walking away from the river's embankment toward the woods. The driver was both startled and curious, so he let off his throttle to slow down. The creature looked to be dull gray in color, like that of a dead tree. As it walked away on two legs, it took huge steps, covering the ground quickly without any noticeable arm movement. As the truck continued to slow, the brakes made a deep rumbling sound. At that point, the driver heard a loud scream coming from the direction of the creature, before it disappeared from view in the darkness. Shaken, the driver accelerated and left the area without stopping in Fouke. It was only later that he was able to do some internet research, finding that others had reported unexplained creatures in the area. Prior to that, he had only associated Sasquatch tales with the Pacific Northwest. Whether anyone else believed his story or not, the driver's mind was permanently changed after what he saw that night.

SWAMP STALKER

By 1999 a few more organizations had joined the hunt for the "Sasquatch of the South," offering additional resources for submitting reports and spreading information. One of these was the Texas Bigfoot Research Center (now known as the Texas Bigfoot Research Conservancy, or TBRC for short), a group of researchers dedicated to proving the possible existence of a large, undocumented primate species in the four state region of Arkansas, Louisiana, Oklahoma, and Texas. The organization was initially founded by Craig Woolheater and Luke Gross, the former having been inspired by the tales of the Lake Worth Monster in Texas, as well as the Fouke Monster.

Since then the TBRC's membership has expanded to include a wide range of skilled personnel with backgrounds in wildlife biology, anthropology, ecology, military intelligence, and law enforcement, just to name a few. Over the years, the group has been consulted on numerous occasions by documentary and television producers seeking input on the mystery. Members have appeared in shows such as *Weird Travels* and *MonsterQuest*, which air frequently on the Travel Channel and History Channel television networks.

Browsing their website's vast database of sighting reports, we find that they too offer several cases from the Miller County area. The first one occurred on January 15, 2000. The witness claims to have seen a "hairy man-like creature approximately eight-foot tall" walking through a wheat field west of Fouke in the late afternoon hours. He described the creature as having thick, reddish brown hair, which was matted with sticks and leaves. The witness also noticed a "distinct musky smell" downwind from the animal.

At one point the alleged creature suddenly stopped and looked at the witness. The man was armed but did not attempt to shoot. In his own words: "I was very frightened when it stopped and looked in my direction. I had a gun but would never shoot anything unless it is game."

After the creature traversed the field, it disappeared into the trees and was not seen again. Perhaps not coincidentally, the field was located on the edge of Boggy Creek.

The next report on file is an even more chilling account, which also occurred during the winter of 2000. The witness in this case was leery about sharing his experience for fear of ridicule, but nonetheless allowed his report to be published online and for a TBRC investigator to follow up with questioning. The result is a detailed account of a frightening face-to-face encounter with a large, unidentified primate.

The incident took place on a cold moonlit night sometime after midnight. The witness and several other men were coon hunting near Mercer Bayou in an area the old timers call Thornton Wells. Coon hunting is typically done using well-trained dogs, so the men in question were using Treeing Walker Coonhounds to locate the target animals. At some point, the witness reported that his dog treed a raccoon on one side of the swamp while the other dogs pursued coons elsewhere in the swampy bottoms. The witness followed his dog's lead and separated from the other men. After locating the coon and collecting his dog, he started back across the swamp to rejoin the group when something strange happened:

> Suddenly I heard something walking in the flooded timber. (Bare in mind this whole area is 8-15 inches flooded with brackish swamp water.) I could tell by the gate [sic] that it was someone coming in the flooded woods. So I called out to them, thinking it must of been one of my hunting companions. To my dismay no one answered—instead all I heard was a deep throated gurgling growl and [smelled] the awfullest [sic] putrid smell. The smell was like when you kill a wild hog and grab him by the hind leg and then you give out dragging him and put your hand up to your nose and the smell knocks you over. I also heard a whining kind of a whistling sound. My dogs were on my leash and were

whimpering. These were UKC Lipper-bred Walkers, they probably weighed in at close to 100 lbs. They could rip a 20 lb coon apart, yet there they were cowardly crouched behind my legs whimpering. I couldn't understand it. Still, in all, I was not concerned about any of this and got my dogs and started back across the swamp.

After a few minutes I heard this sound of someone or something walking in the water again. So I stopped and turned around and standing right behind me was a creature of immense size. Talk about the hair on the back of your neck standing up, I was scared. I don't ever remember being that scared. He made a hissing sound and reached down and took his hand and started scooping water and throwing it up at me and making a deep throaty noise. My dogs started chomping at the leash and growling trying to get at it, having regained their bravery. I grabbed the leash and tore out across that swamp so scared that, for awhile, I went in the wrong direction. I got to a big Cypress knee and caught my breath. I could hear the thing behind me, it sounded like it was about a 100 yards or so back. It followed me for a ways, then after a while I could hear it across the bayou making a moaning sound and moving away slowly.

I got my breath and my compass and my bearings and started back to the truck. When I got there I didn't say a word to my friends about what I saw and heard for fear they would laugh me down. Needless to say I never returned to Thornton Wells nor do I plan to. For a long time I kept this to myself. I even had nightmares about it.

A follow-up to the report was made by Daryl Colyer, a long-time TBRC investigator with a background in military intelligence. After interviewing the witness, his assessment was that the man seemed very credible and would have no reason to lie about what he had seen. Colyer's report also brings out additional details:

> Something of note that struck me about the incident was the splashing of the water by the subject as the witness and his dogs attempted to assess what they were seeing. Apparently, in an attempt to chase away the

witness and his Walker coon dogs, the unknown subject splashed water with its hand toward them. The witness said he could clearly see its hand, which he described as looking like a "huge monkey hand." The witness added that the skin was visible on the hand and that it was "gray and leathery."

The face was also a "grayish color" with no or very little hair on it. The hair on the head appeared "slicked back" and the subject in general appeared "wet." The overall coloration was "dark, rusty brown" or a "similar but a little darker than pine needles" after they have fallen on the ground and turned color. The witness said he did see a "flash of teeth" and they appeared human but much larger. The witness apparently was haunted because the face, as he described it, was "like that of a person."

I found the reaction of the coon dogs of interest as well. Initially, before the witness established eye contact with the subject, the dogs were "nervous" and "whiney," and stayed behind the witness. Upon seeing the subject, the dogs became emboldened and it was all the witness could do to keep the dogs from breaking free from their leashes to attack. The witness felt that the dogs may have saved his life.

Given the length of the sighting, it is hard to believe that an experienced hunter such as this would mistake any sort of common animal for that of a large, human-like primate. If the story is true, it firmly establishes that something fitting the description of the Fouke Monster was still roaming the swampy bottomlands of southwest Arkansas in the 21st century.

The next report from 2000 was also submitted to the Texas Bigfoot Research Conservancy and thoroughly investigated by a lead TBRC investigator who found it credible. The witness this time was an avid bowhunter by the name of Stacy Hudson, who participated in a recreation of the event for the *MonsterQuest* "Swamp Stalker" episode.

It was a cool, crisp morning in October 2000. Hudson had set up a deer stand in the vast Sulphur River Wildlife Management area less than ten miles from Fouke. He was settling in to wait for deer as the full moon played out its last few minutes of life. In his own words:

> I was up the tree and had my bow in my lap with my arrow pointing away from me with it resting on my stand's gun rest. I pulled my earth scent wafers out and attached them to the outside of my back pack.
>
> I was facing south with two big oak trees about 25 yards away. The clouds were moving fast and with the full moon, I could somewhat see in the dark.
>
> I was very still and had everything camouflaged except my muzzy broadhead, on the end of my arrow. It was about 5:30 a.m. when I heard something walk up from behind me.
>
> I turned slowly to see if I could see it. It was big and black and at first thought it was a couple [of] hogs. They were too big to be hogs and I wondered whose cows were loose in the refuge. I was wanting them to hurry up and leave so they wouldn't scare the deer when something brushed up against my leg. I didn't hear anything climb up the tree with me so I kept still. I was looking forward when I saw a big black hand reach up and grab my muzzy broadhead. The hand was a shiny black and the fingers were huge! The palms were also black like a gorilla. When it grabbed my broadhead the razor blades cut him and it yelled like a bull! It was louder than a car horn and it was at my feet. It scared the crap out of me so I started yelling back.
>
> It ran off breaking trees and tearing down everything in its path. I sat there 15 minutes until it got daylight and left. I believe that I scared him/her as much as it scared me.

The report was investigated by TBRC member Jerry Hestand. According to Hestand:

> I had the opportunity to interview this witness, an extremely experienced outdoorsman and bowhunter, several times over a period of a few months. His

2000: A coonhunter has a frightening experience in Mercer Bayou near Thornton Wells.

statement never altered and I believe he was truthful at
all times.

The incident was an event that changed his way
of thinking for sure, even though he had already
experienced some incidents that were highly unusual.

Another incident worth noting occurred sometime between
2000 and 2001. At the time, Denny Roberts (brother of Rick Roberts)
was at the Monster Mart convenience store on Fouke's Main Street.
One afternoon he spoke to a man who was doing survey work for
the Army Corps of Engineers as they prepared the site for the new
Highway 549 construction project. According to the man, he had
been working in a wooded area somewhere south of Fouke when
he came upon the remains of a large hairy animal. The carcass was
already starting to decay, giving off a foul odor, but still he could
tell that it had been roughly the size of a large human and covered
with a layer of dark, matted fur. It did not appear to be a bear or
any familiar animal, but there was nothing he could really do at the
time to further determine what it might be. According to Denny, the
man claimed that when he returned to the site the following day to
continue his work, the carcass was gone. Apparently the man was not
familiar with the stories of the Fouke Monster, as the subject only
came up while discussing strange things he had seen while working
in the area.

Another sighting in the same area was reported in town by a
visiting hunter. In this case, he claimed to have come face-to-face
with a large ape-like creature while hunting near Mercer Lake. He
was so frightened by the encounter that he dropped his expensive
rifle and ran. He refused to return to the area, even to look for his
gun. The incident occurred near the old train trestle that spans
across the Sulphur River just west of Highway 71. Remnants of the
old structure still remain today, although the railway has long since
been removed.

While showing some out-of-town visitors around one day, Rick
Roberts brought them to the location, pointing out that it had been
the site of several eyewitness reports. As they approached the area by
car, exiting off of the highway, they noticed a dark figure standing in

the trees just east of the trestle columns. Its movements did not seem consistent with that of a bear, so they tried to get a better look. But by the time they got closer to investigate, the figure was gone.

Remnants of the old train trestles spanning across the Sulphur River.
(Photo by the author)

Speaking again to Sheriff H.L. Phillips, I learned of another incident that involved people not previously acquainted with the legendary beast. According to Phillips, he took a report in 2003 from a family in Jonesville who said their two children had been out riding bikes when they came up on what they described as a "hairy monster" during the evening hours. The kids were quite shaken and quickly peddled away on their bikes. When they returned home and frantically told their mother, she decided to contact the authorities. The Sheriff's office took the report seriously, as they always did, and asked if she thought the children had seen the Fouke Monster. The woman was puzzled. They were new to the area, having only lived there a short time.

Neither she nor the kids had ever heard of the Fouke Monster.

2003: Still haunting Jonesville—the creature is seen by two children.

7. A Question of Theories

Man-Made Man-Apes

The culture surrounding the alleged creature is interesting in and of itself—the excitement of the sightings, the entertaining movies, the dramatic documentaries—but still it leaves one wondering if there is any truth or reality to this whole business. Could the Fouke Monster be real?

Over the years, various explanations have been offered up by locals and outsiders alike in an attempt to explain away the mystery. Stories involving derailed circus trains, moonshiner conspiracies, rogue panthers, or a person in a suit have been spread around like urban legends—or in this case *rural* legends—until no one is really sure how most of them got started. While some of these theories may have merit, most don't come close to offering a suitable explanation for the entire phenomenon.

Beyond these mundane explanations, there is the glaring question of whether or not a large animal, such as the one described in these events, could even survive in the immediate area of Fouke. Could a large, man-like ape really remain hidden in the Sulphur River Bottoms without being captured? Short of capturing one or finding an actual body, we may never know, but at least by addressing some of the long-standing myths, we can come closer to a satisfactory conclusion.

Before evaluating specific theories, it is worth addressing the general notion that only rednecks and crazy people tend to see such "monsters." While cabin fever or mental issues could certainly cause a person to see something that is not there, this "explanation" is basically nonsense. While people living in rural areas may not possess the refined communication skills of college graduates, the

161

fact is that city folks are unlikely to spot such a creature while strolling the alleyways of their concrete jungles! Rural folk are simply far more likely to witness a strange wild animal than someone who does not live near a significant patch of woods or other remote area. So it is not surprisingly that most sightings of the Fouke Monster have been made by country folk and hunters. This fact doesn't validate the Fouke Monster, but at least it should not raise a red flag, as popular culture might suggest.

As for the question of "crazy," I have not found this to be a predominant characteristic of people who claim to have sightings. While I have come across some folks whose stories or demeanor have led me to believe that they are either seeing things or making things up, those are the rare exception and their stories were not considered for this book. On the contrary, the majority of people I have talked to who claim to have had sightings of Sasquatch-type creatures seem to be of sound mind. Beyond the average down-to-earth hunter and country folk types, the witness demographic ranges widely. Some have been college graduates, college professors, medical researchers, military personnel, police officers, truck drivers, and other upstanding professions, which shatters the feeble "crazy redneck" theory the media tends to favor. That might make for good television and shield the reporters from making any serious conclusions on the air, but in reality some very sane and credible people have reported seeing hairy, ape-like creatures lurking in the woods of Arkansas, Texas, Oklahoma, and surrounding states. Again, this in and of itself doesn't prove anything, but it does force one to consider the phenomenon more closely.

BLAME THE TRAIN

The first theory that usually comes up in conversation is that of the so-called "circus train wreck." In this scenario, which is by no means exclusive to the Fouke Monster, the sightings are attributed to a circus animal, namely a primate, that has survived the wreck and promptly taken up residence in the nearby woods. While circus accidents have been known to happen, there is little evidence that these events occur with any regularity, much less in the areas in question. Nor could they possibly account for all the strange sightings that people have been reporting. However, since most of the Fouke Monster sightings tend to describe the creature in terms of a primate, it seemed worthwhile to research the possibility that some sort of circus calamity is at the root of the phenomenon.

With a little research, I learned that the worst circus train accident in U.S. history occurred on June 22, 1918, in Ivanhoe, Indiana. The Hagenbeck-Wallace Circus train was parked on a side rail after making an emergency repair stop, when it was struck by an empty troop train whose conductor had fallen asleep at the throttle. Of the 300 passengers onboard the circus train, 86 were killed and 127 were injured. The number of animal fatalities were not reported.

Another famous wreck involving a circus train occurred on August 6, 1903, in Durand, Michigan. In this instance, two trains transporting the Great Wallace Brothers Circus crashed into each other when one train's brakes failed. Some animals were killed, but human travelers took the brunt of the collision with a total of 23 people dead.

So, while these type of incidents do happen, it does not come close to explaining anything in Fouke, unless one of the displaced primates happened to catch an outbound train for Arkansas. Fouke resident Willie Smith stated once that he believed the Fouke Monster was an animal that had been left behind by a circus or traveling show. "It could have escaped," he told the *Texarkana Gazette* back in 1971, "or maybe someone let it loose." Smith's daughter-in-law, Lynn, even cited a date for an alleged circus crash. She claimed that a "traveler coming through Fouke mentioned to her that about 1953 a wreck of a circus truck reportedly occurred and some wild animals escaped."

Armed with an approximate date and the tidbit about it being a *truck* not a train, I dug deep into news archives until I found something that made my jaw drop. A circus transport had indeed crashed in Arkansas around this time. It was not in 1953, but in 1951. And on the list of escapees were three monkeys!

My heart beat rapidly as I flipped to the article. Could this reveal a zoological smoking gun? Reading the short blurb, I found that the crash did not occur in the Fouke area, but in Mena, Arkansas, more than 100 miles away. Nonetheless, it was still an interesting fact. The blurb appeared as part of a larger article titled "Animals: Battle of Species" which ran in the November 12, 1951 issue of *Time* magazine:

> In Mena, Ark., while the Campa Bros. circus trumpeted through its one-night stand, nine-year-old Maria Campa, granddaughter of one of the circus owners, was clawed and chewed to death by a young lion considered so tame he was tied to a stake outside his cage. Next day, as the Campa circus trundled along the rain-slicked road toward Mount Ida, two trucks overturned. Nine beasts scampered into Ouachita National Forest. A pursuing posse brought down one of two escaped leopards and recaptured a tame black bear and a rhesus monkey. The other leopard prowled all night before being tracked down by a small but heroic cur named Tony, whose owner, Roiston Fair, shot the leopard, but not before it had killed Tony. Still in the forest: a polar bear, a black bear, three monkeys.

Mena, Arkansas sits about 120 miles north of Fouke on Highway 71. It is unlikely that the escaped monkeys made their way all the way to Fouke from the Ouachita Mountains,[15] but nonetheless, this seems to confirm the rumor heard by the Smith family, which claimed that a circus truck had crashed in the region. Perhaps the traveler mentioned by Mrs. Smith had confused the two areas. If he was not from Arkansas, he might not have realized how far Mena was from Fouke.

I continued to dig deeper, trying to determine if another such crash had occurred closer to Fouke. In searching various archives, including www.circushistory.org, I was able to verify that a number of circuses played Texarkana in the years prior to and after the Jonesville/Fouke Monster began to surface in the 1960s. Circus troops such as Ringling Brothers and Barnum & Bailey, Cole

Two Fouke kids display an alligator pulled from the Sulphur River.
(Courtesy of the Miller County Historical Society)

Brothers, Downie Brothers, King Brothers, and Mills Brothers regularly played in Texarkana, including dates in the year of 1953. Many of their touring routes went from Texarkana to Shreveport or vice versa, which would have taken them right up Highway 71 where the monster was known to haunt.

Though the details of these routes are intriguing, I failed to uncover another circus wreck that happened in the vicinity of Fouke. In the end, I could say with a fair amount of certainty that circus-related tragedies, although rare, are typically well documented. So it stands to reason that if something like this had occurred in the Texarkana/Fouke area, I should have been able to find some record of the event.

I also discussed the circus train mystery with Miller County historian Frank McFerrin, who has written extensively about the history of railroads in the area. He had also heard of the train wreck theory but had never come across any evidence that could explain why non-indigenous apes might be loose in the area. McFerrin and a few other locals seemed fairly certain that a cargo train had derailed at the location where Boggy Creek enters Days Creek, but there was nothing to suggest it was carrying any circus animals.

Given the sketchy details of this derailment and the facts about the circus truck crash in the Ouachita Mountains in 1953, it is likely that the circus train myth developed out of one of these incidents or a combination thereof. Something else may have happened that was never recorded, but there's no way to be certain. It does seem that an event of this nature, if it occurred on Highway 71, would have been quite newsworthy and would have received at least some coverage in the *Texarkana Gazette*. Nearly every brief sighting of the monster garnered attention, so it stands to reason that an overturned circus truck spilling out a family of apes into the nearby woods would be major news.

This being said, I am not ruling out the potential presence of circus-worthy animals who managed to visit Fouke by other means. It is a fact that a full-grown African lion was loose in the area around 2008. Most likely it was an exotic pet that had escaped into the heavily wooded area near Jonesville. On one occasion, Rick Roberts and his son saw the lion standing in an open field about 50 yards from one of the old county roads near Jonesville. It turned up months later on the other side of Highway 71, severely famished and covered in blood, as reported by a female witness. Nothing else is known about this strange visitor, but either way, it is not likely that it was ever mistaken for the Fouke Monster.

The example of the wild lion does at least prove that it's possible for exotic animals to find their way into rural settings. If an orangutan, chimpanzee, or gorilla managed to escape its owner and run off to the woods down by Fouke, then perhaps it could be mistaken for a monster when seen on a moonlit night or from a considerable distance. But this does not account for the descriptions that say the creature walked upright like a man or that it was running on two legs. Nor does it take into account that most people, even good ol' country folk, are familiar with common apes and would be able to discern them from "hairy monsters." If the Fouke Monster were an escaped pet, then most likely it would end up being properly identified or caught. But this never happened.

MOONSHINE MASTER PLAN

Another theory that comes up in conversation is the "moonshine monster." This theory suggests that perhaps the Fouke Monster was created to keep people away from illegal whiskey operations. While there were definitely moonshine stills operating in and around Fouke, there is little evidence to suggest that moonshiners invented a monster for such purposes.

Stills existed in Arkansas prior to prohibition in 1920, but it was this particular American fiasco that kicked up demand and got the alcohol really flowing, so to speak. Out-of-the-way rural locations such as Fouke were ideal for hiding illegal stills, and indeed these type of hidden production facilities boomed in the area. They became so rampant, that in 1931 the resident Federal officer in Fouke, Jess Quillen, claimed it was a "war" after a local liquor raid resulted in the death of Miller County Sheriff Walter Harris.

Following his death and a renewed vow to bring the bootleggers to justice, hard-nosed cop Rufus Turkette was appointed as the new Miller County Sheriff. Within six months of Turkette being in office, an astounding 89 stills were destroyed and numerous bootleggers arrested in the process, many of them in the immediate Fouke area.

With the end of prohibition in 1933, things calmed down but this did not completely eradicate the illegal liquor business. The demands of prohibition had created a new breed of moonshiner who was not going to stop simply because liquor was again legal. Keeping it underground kept them from paying taxes on their product. As a result, stills remained active in the area, which attracted an ever increasing flow of criminals, including Charles "Pretty Boy" Floyd and the infamous duo of Bonnie and Clyde. One celebrity appearance is detailed in *Fouke Arkansas In Word and Pictures 1891-1941* (1991):

The prohibition era and the great Depression produced an environment in which crime flourished. Several heavyweight criminals and their associates came into the area.

Butler Pruitt, who worked with J.W. "Bill" Parker in the logging business went to Parker's house one day near Sulphur River for a visit. When Pruitt arrived he noticed that Bill already had guests, a well dressed young man and woman, who seemed friendly. Later he was told by Parker that the visitors were Bonnie Parker and Clyde Barrow and that they had come into the area to visit with friends and rest for a day or two. It was alleged that Bonnie was kin.

Illegal liquor operations continued long after the era of Bonnie and Clyde, but they were less prominent as a source of major crime. Yet, even into the 1960s, nearby stills continued to be raided. According to an article in the April 19, 1964, edition of the *Texarkana Gazette*, federal liquor agents raided two moonshine liquor stills that week in Stamps, Arkansas. Many more continued to operate under the radar in and around the Sulphur River area, and you might even find a few today. (But I don't recommend you go looking!)

So, while there is no doubt that the Fouke area has a long history of moonshine activity, what evidence is there to suggest that moonshiners were responsible for the monster tales? Like the circus train theory, it seems that everyone I spoke to while doing research in Fouke knew of the moonshine theory, yet no one could pinpoint a single confession or conversation that links this theory to reality. Fouke/Jonesville is a very tight-knit community, and like other small towns, people are intimately familiar with one another. When someone speaks of the monster, word gets around. If an old moonshiner claimed that he or she had created the monster, word would have gotten around. Yet, not one person, not even the old timers, can cite a single person who has claimed responsibility for the monster's creation.

But just for argument's sake, lets say that the monster *was* a masterful myth purported by a group of clever liquor entrepreneurs. Would they spread stories of imaginary sightings or convince others to claim that they had seen a big hairy monster lurking outside their window or run by their car at night? To do so would be completely contrary to the low profile most early witnesses tried to keep for fear of ridicule, or worse, a trip to the loony bin. As we have seen, most of the early tales came to light as a result of other people spreading the word, such as news reporters or filmmakers, not through the deliberate spreading of rumors by the people who had seen the monster. The purposeful moonshiner theory cannot account for the sightings by younger kids like Lynn Crabtree, Kenneth Dyas, or by hunters such as Phyllis Brown who kept quiet about their sightings for many years before their stories leaked out. And what about Bobby Ford who had only lived in the area for a few weeks? How could he have so quickly conspired with moonshiners on their master plan. It's just not possible.

Furthermore, to complete the farce, it would require someone to dress up in a fur suit and take a midnight jog down a dark stretch of Highway 71, as in the case of musician Carl Finch's sighting in 1967. Finch was an outsider, having no prior knowledge of "the monster," so it seems incomprehensible that these type of sightings could have been staged by the moonshiners. It just doesn't add up.

Again, as in the case of the circus train, this does not completely rule out the possibility that a moonshiner or two didn't try to use the monster to their advantage, but it cannot account for all the sightings over the last 60 years. Besides, if this was a masterful plan by moonshiners, it totally backfired. On more than one occasion, the monster incited hundreds of people to take to the local woods on a monster hunt. That's not the kind of thing you want to happen when you are trying to kick back at the hideout and enjoy a jar of White Lightning with a few close friends.

On the flipside, some have claimed the monster owes his creation to the fevered visions of those who were *customers* of the moonshiners. While this may account for a sighting or two, it certainly cannot account for all of them, or even a majority. Many of the sightings were reported by respectable people, skilled hunters, and policemen, who were definitely not drunk at the time. To them, the monster was a sober reality.

THERE'S A PANTHER UNDER THE HOUSE

The next theory up for evaluation is the "panther" explanation. In this scenario we are asked to believe that a large cat, specifically a *black panther*, is responsible for some or all of the sightings and track finds.

In order to examine this more closely, we must first decide if any so-called panthers live in the area around Fouke. The answer should be fairly easy to come by, but for some reason most regional wildlife management agencies like to play down the existence of predatory felines, often denying their presence altogether, despite overwhelming eyewitness reports. Ironically, this sounds very much like the dilemma of the Fouke Monster.

The term "black panther" is itself confusing since it is not a distinct species of cat, but rather a generic term for black (melanistic) specimens of the genus. In other words, "black panther" is just a commonly used name for the dark colored version of *any* species of large cat. For example, black panthers in Latin America are actually black jaguars (*Panthera onca*), and in Asia or Africa they are black leopards (*Panthera pardus*). In North America, they may be black jaguars or possibly black cougars (*Puma concolor*), although no cougars have ever been proven to have this variant.

Confused yet? No problem. Just keep in mind that a "black panther" is not necessarily one specific cat, but simply *any* large cat that is black in color. Beyond that, the cougar, also referred to as a mountain lion or puma, is primarily tan or gray in color with a lighter underbelly. No doubt this type of cougar comes to mind when one thinks of North American wild cats.

All that being said, do any *large cats of any color* exist in Arkansas? Well, yes and no. Mountain lions (cougars) were known to be present in Arkansas until their apparent eradication, which most agree occurred by 1920. In 2001, the Arkansas Game and Fish Commission took mountain lions off the state's endangered species list and officially adopted the policy that there are no wild mountain lions in Arkansas. However, some sightings have been reported and the animals have been photographed, indicating that these animals still roam free in the state, even if only in small numbers.

An article which appeared in the December 22, 2006, edition of *The Daily Citizen* out of Searcy, Arkansas, contained the details of a sighting four miles north of Searcy. A mountain lion had been seen on several occasions by multiple people. The *Arkansas Democrat-Gazette* cited an incident in August of 2003, when a hunter's game camera captured the image of a mountain lion on private land near the Winona Wildlife Management Area west of Little Rock. That same year, the *Democrat-Gazette* ran another article detailing a study by the biology department at the University of Arkansas at Little Rock, which serves as the state's clearinghouse for cougar reports. The study was headed by David W. Clark, a former UA graduate student, who stated: "We have documented a minimum of four mountain lions in Arkansas over a span of five years based on Class I evidence." This evidence included photographs, scat, tracks, and casts of tracks.

The Game and Fish Commission attributes these rare cases to escaped pets rather than to remnants of the state's original mountain lion population. According to Game and Fish biologist Blake Sasse: "We haven't come across any [mountain lions] in Arkansas that we can't trace back to a pet animal that's escaped or intentionally been released."

But regardless of how the cats got there, they have definitely been there, and were just as likely to have been present back in the mid-1900s when the whole Fouke Monster business got underway. As for the black panther variation that some have claimed is responsible for some of the monster incidents, that's even more debatable; although, again, there have been reports of melanistic mystery felines all over North America, including the state of Arkansas. In fact, one of them was clearly spotted in the Texarkana area in June of 1971, during the peak of Fouke Monster activity. While Deputy H.L. Phillips and Sheriff's Posse member Gary Beard were patrolling near Sugar Hill Road, several hundred yards from Highway 71, they saw what Phillips referred to as a "darn big black panther." It walked right in front of the patrol car, so there was no doubt about what they had seen. Phillips and others conjectured that it was this creature, or another like it, that may have been responsible for some of the Fouke Monster incidents.

Resident Rick Roberts has personally seen two black panthers near Fouke. The first was at Coon's Crossing near Jonesville; the second was seen by both Rick and his brother Denny as it lurked around Denny's property in Fouke.

I also had a rather timely sighting of a strange feline on one occasion as I was leaving Fouke. My wife and I had spent the day conducting interviews and doing various research for this book and were heading home just before dusk. We traveled south on Highway 71 from Fouke and turned on the access road leading to the newer Highway 549. As we began to merge onto 549, my wife pointed to an animal standing in the grassy area off to the side of the shoulder. We were moving quickly, but there was no doubt at all that it was a black cat. A very large black cat. As we passed by and picked up speed, I turned and looked back so I could watch the animal for as long as possible. It did not appear as large as a full grown wild cat, such as you would typically see at a zoo, but it was noticeably larger than a common house cat and there was no doubt that its fur was a uniform, solid black color. After it faded from view, I turned back and we discussed what we had both seen. There is no way to be certain, but we both found it ironic to see what we believed to be some kind of medium sized black "panther" lurking right there on the outskirts of Fouke.

So now that we know big cats "exist without existing" in Arkansas, could they possibly be masquerading as seven-foot-tall man-like apes? Going back to the Ford incident, there is some evidence that suggests the family's mysterious visitor may have actually been some type of panther, black or otherwise. Upon the initial investigation conducted by Constable Ernest Walraven, he found evidence that something had torn away some of the tin that was nailed around the bottom of the house. The house sat about three feet up off the ground, so there was ample room under it for animals to hide. He also observed scratch marks and tracks that might indicate some kind of large cat had been making a home under the house. Even Don Ford was quoted a few days later as saying, "We think now it might have been a big cat, like a mountain lion or puma."

No further evidence had surfaced to show that it was such a cat, so the reason why he changed the story is not clear. Maybe it was due to the hundreds of people who were crawling all over the property on the Monday after the news report came out, making it hard for him to sleep. "I work nights and haven't been able to get any sleep today," he remarked to newspaper reporters. By minimizing the monster, perhaps he thought the hoopla would die down.

In Bobby Ford's initial statement, he was quoted as saying: "I was walking the rungs of a ladder to get up on the porch when the thing grabbed me. I felt a hairy arm come over my shoulder and the next thing I knew we were on the ground. The thing was breathing real hard and his eyes were about the size of a half-dollar and real red." Presumably, it was the same hairy arm that had poked its way through their window on the previous night. It's hard to imagine that they would not recognize a cat's paw, even if it was black in color, but perhaps it could be mistaken for the hand of some other more *mysterious* creature. Even in Bobby's struggle to reach the porch, it was dark outside and perhaps it was too difficult or he was too frightened to focus on the creature long enough to get a better look. He felt that he was fighting for his life, so that would be understandable.

Tracks found on subsequent occasions, such as the ones spotted in June of 1971, near the abandoned fertilizer plant on Oats Road, after sightings were reported in the area, could also be attributed to such an animal. However, there was no direct link to the Fouke Monster, so even if the tracks were made by a panther, it did not implicate the monster as a mere cat.

So in the long run, even if we credit a black panther for the attack on the Fords, for some of the track finds, and maybe even for another vague sighting or two, this doesn't come close to a satisfactory explanation for the sightings as a whole. Like the moonshiner or hoax theories, the black panther theory does not account for why highly seasoned hunters saw hairy man-like animals *walking upright* in the woods, or running by their tree stands, or why countless people had clearly seen a bipedal creature resembling an ape walk across roads in plain view.

The black panther is indeed a high ranking player in the world of cryptozoology, but his story is not that of the Fouke Monster.

SUNDOWN TOWN

Another theory to explain the Fouke monster is rooted in one of the darkest corners of American history. This particular rubbish theorizes that the monster was invented as a means to keep non-whites out of Fouke. While it is not something that most of us are proud of, it's a fact that racism has plagued our country since the very beginning. After the abolishment of slavery in 1865, things got even worse, causing scenes of racial tension and violence to be played out all across the U.S. Every facet of our society has been affected, and indeed it seems that few cities or towns were immune to the issue. Fouke was no exception and even found itself on a list of suspected "Sundown Towns."

The term "Sundown Town" is used to identify towns or cities that are suspected of resisting integration, specifically when it comes to African Americans. One basis for being labeled a Sundown Town is the number of African American, Hispanic, or other races living in the town versus the number of Caucasians. Another basis is whether any racist laws were ever on the town's law books. In the case of Fouke, no such formal laws existed, but there is no denying that the town has always been populated by an almost exclusive Caucasian demographic and was resistant to integration at some point in the past. This being so, whispers spread that "blacks" were not welcome and, at times, outsiders joked that the Fouke Monster was created to keep it that way.

To make matters worse, an old Pepsi promotional item became an unintentional player in the Sundown saga when it was offered to Fouke as a "Grand Prize." Sometime in the 1940s, a painted statue of a black man pointing into the air was bestowed upon the town. Its intention was to point at a nearby sign which advertised Pepsi-Cola for five cents. However, at some point someone turned the statue so that it seemed to point out of town. Whether intentional or not, this was interpreted as saying "blacks turn around and leave," or as the name Sundown Town suggests, "... don't let the sun set on you in this town... or else."

The Pepsi-Cola promotional statue pictured sometime in the 1940s.
(Courtesy of the Miller County Historical Society)

Regretful as it was, the statue remained that way until the 1960s when one local man finally had enough. Under the cover of night, he backed his truck up to the figure, threw it into the bed of his truck, and drove off. He then made his way down to the Sulphur River where he heaved it into the muddy water. The out-dated promotional statue sunk to the bottom, never to be seen again.

The statue and its all-white population did not reflect well on the town, so that when the civil rights movement gained momentum in the late 1960s, right about the time the Jonesville/Fouke Monster sightings were becoming more frequent, it was inevitable that some would theorize that the town may have concocted the monster as a warning to minority outsiders. But like the moonshiner theory, there is little to suggest that these two elements had any connection whatsoever. First, the town was not being threatened by an influx of minority visitors or residents, so it would seem unlikely that it would need something as far fetched as a monster to keep it that way. The statue had already proved to be an uncomfortable icon. What more could a fictional monster do?

And just as in the moonshine case, the reluctance by the witnesses to spread their tales suggests that this would be contrary to the desired effect of spreading the monster story as far and wide as possible. How would minorities be "warned" unless the word was spread to the outside world? Many Fouke residents I spoke to had heard this particular theory, yet again, no real source could be traced back to confirm it.

The facts surrounding the Sundown Town theory are not pretty, but this was a dark facet of life during America's formative years, not just for Fouke, but for other places as well. The rationale behind segregation seems old fashioned and cruel now, but in those days it was a part of the culture.

HIDDEN HOMINOID

The last major theory holds that the Fouke Monster is some sort of unknown biological entity, specifically a large bipedal primate that has existed beyond the reach of modern science. In simple terms: a Sasquatch-type creature.

As we all know, circumstantial bits of Bigfoot evidence, such as blurry videos, sighting reports, or footprint casts, have been scrutinized for years by numerous cryptozoologists, scientists, amateur researchers, and television shows, but as of yet, there is no definitive proof one way or another. The various theories regarding Bigfoot's existence are wide ranging and have been the subject of countless books, but in the interest of the Fouke Monster, we will consider only that which has come to light in this case. If such a creature does exist, it may ultimately be linked to a larger family of unlisted primates nationwide, but for now it will be treated on its own merits as a creature indigenous to the four state area where it appears to dwell.

In this regard, to be a flesh and blood animal, it would require proper habitat, food, water, and a breeding population. There could not be a single "Fouke Monster." A family is required. The creatures would also require a suitable habitat where they could thrive, a place that would provide plenty of cover far from the eyes of humanity, otherwise they would not be able to endure so long without discovery.

There is no doubt people in the area have seen an entity that they believe to be a seven-foot hairy man-like primate, so it can only be assumed that on occasion the creatures are willing to venture from their swampy stronghold closer to the boundaries of civilization. It is on those occasions that they come into contact with some lucky—or perhaps unlucky—witness, as the case may be.

Could such creatures even exist near Fouke? If the area is not remote enough or could not support a population of large creatures, then the validity of this final theory is diminished right out of the gate. Fortunately, however, it seems that if our mystery beast is indeed a living, breathing hominoid, then it has chosen a likely stomping ground. In fact, it couldn't be more ideal.

8. The Burden of Proof

Land of the Southern Sasquatch

"He always travels the creeks." So say the locals. And certainly there are plenty of muddy tributaries, leading from the deep reaches of the Arkansas bottomlands to the edges of civilization, that the alleged creature could follow. In fact, if we were to go looking for a such a thing as a "Southern Sasquatch," this would be a logical place to start. The area is not only a marshy bottomland, but a densely forested region that has all the necessary ingredients to host an animal like the one reportedly seen near Fouke.

As I mentioned previously, Boggy Creek, Days Creek, and the other bodies of water in the area are considered part of the Sulphur River Basin or Sulphur River "Watershed." This great network of waterways provides some of the best habitat for supporting natural wildlife in the southern states area, and perhaps not surprisingly, Fouke sits almost smack dab in the middle of the richest concentration of forestland in the four state region of Arkansas, Louisiana, Oklahoma, and Texas. By combining the total forestland at the intersection of these four states, which is only miles from Fouke, we arrive at a staggering 65 million acres, or 100,000 square miles.

Investigators Daryl Colyer and Alton Higgins of the Texas Bigfoot Research Conservancy have built a convincing case that the forestlands and bayous in question would be ideal for harboring the target species:

> While the forestlands of Texas, Arkansas, Louisiana and Oklahoma may be somewhat more parceled, or discontinuous, than northwestern forests, it is obvious

that they are enormous in scope and depth, contrary to the misperceptions of some. Wildlife biologist Dr. John Bindernagel, who visited the region in 2001 and 2002, was struck by the richness and scope of the region's forests, which are predominantly mixed deciduous, as opposed to the largely coniferous forests of the Pacific Northwest. Dr. Bindernagel recognized the value and productivity of deciduous forests in terms of wildlife habitat and he pointed out that large species of mammals living in the southern forests would almost certainly require smaller home ranges than in northern coniferous forests.

Almost without exception, reported Sasquatch sightings occur near water. This is even true with the relatively few reports originating in the drier regions of Texas and Oklahoma, where Sasquatches are reportedly seen generally on or near waterways or lakes in thick brush or dense riparian vegetation. Most wildlife researchers and hunters would quickly reinforce the observation that many mammalian species often use rivers and creeks as travel routes. Since water is essential for the cycle of life, animals regularly congregate near or at least dwell primarily in areas featuring bodies of fresh water. Both Texas and Oklahoma have an abundance of rivers, creeks, swamps, reservoirs and lakes, particularly in their eastern regions. It is also reasonable for a large number of reported sightings to occur in or around swamps, river bottoms or bayous, since a reclusive, shy animal would find seclusion and sanctuary in such areas.

When a river basins map is viewed with an overlay of reported encounters and an annual rainfall overlay, it becomes evident that most alleged sightings have occurred along waterways and lakes and in areas with thirty-five inches or more of annual rainfall. Many reported sightings in Northeast Texas have occurred in the Red River Basin along the Sulphur River or Red River and/or their adjoining reservoirs or creeks. Many reported encounters have also occurred in the Red/Sulphur River watershed in southeastern Oklahoma, southwestern Arkansas and Northeast Texas.

Given the data, it would appear that this area—perhaps more than most—would be capable of supporting a small population of large hairy mammals. The vast region of riparian woodlands and river networks, especially those that spread south from Fouke, would provide plenty of places to live, eat, and hide far from the eyes of humanity. It would only be on those rare occasions when the creatures ventured out of the bottomlands, that they could be seen in shadowy glimpses. Famed Sasquatch researcher John Green of British Columbia tends to agree with this scenario. In his definitive book, *Sasquatch: The Apes Among Us*, he writes: "If the bottomlands do indeed provide the habitat for hairy bipeds it is not at all necessary that all the sighting should take place there. It would be when they ventured out into the inhabited territory or crossed roads that they would be most likely to be seen, and that is in accord with the reports."

Some theorize that the creature's occasional ventures into more inhabited areas can be attributed to a rise in water level. During times of heavy rainfall or runoff, the bottomlands can become flooded, thereby forcing the creature to retreat to higher ground. This, in some cases, may result in the creature having to tread closer to civilization than it normally prefers. Several people who have lived in Fouke all of their life, and followed the creature's sightings, told me that they have seen a correlation between the amount of seasonal rainfall and the number of sightings.

Since the area of prime habitat extends well beyond Fouke, it would stand to reason that creatures fitting the general description of the Fouke Monster are likely to be spotted throughout the four-state region. And that is precisely the case. As we learned earlier, stories of "wild men" in the area date back to the 1800s. In modern times, eyewitness accounts of mysterious ape-like beasts continue to be reported throughout the region with a staggering consistency. Encounters have been reported in more than a dozen Arkansas counties, including the nearby Lafayette and Little River counties, as well as others across the state such as Union, Ouachita, Logan, Montgomery, Saline, Baxter, Benton, and Hot Springs.

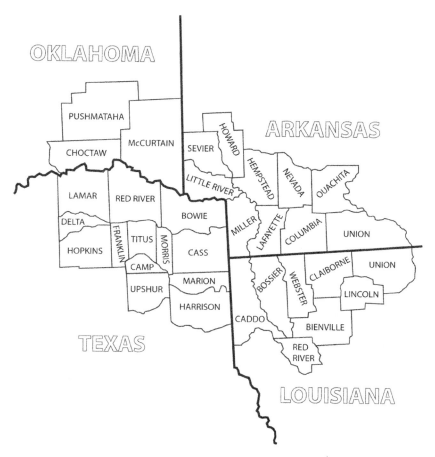

Counties near Fouke, Arkansas

In Little River County, just north of Miller County, several encounters with large, hairy bipeds have been reported along the Red River. In 1969, near the town of Alleene, several witnesses reported seeing an unexplained creature while on a camping trip. One was seen when a female camper got up during the night to relieve herself. As she walked from the tent, she was frightened by a large "half man, half ape looking creature" watching her from the edge of the trees. The creature was standing like a human with a slightly stooped posture, in full view approximately 15-20 feet away. It towered an estimated seven feet tall and had 4-6 inch fur on the

1969: A woman encounters a half-man, half-ape creature while camping near Alleene, Arkansas.

torso with shorter fur around the head and face. Its eyes seemed to glow with an eerie red bioluminescence.

Horrified, the girl ran back toward the tent. In doing so, the creature was apparently startled as well, and in making its own escape, ran full tilt into the side of their pickup truck before running off into the trees. The noise woke her father and brothers, who decided it would be best to pack up and leave immediately. As they packed hurriedly, they saw the creature crossing a nearby logging road, apparently curious to see what the people were doing. They noted that it moved rapidly on two legs and never dropped down to all fours. Before leaving, the men found footprints where it had run into their truck. The witnesses stated that the prints were clearly defined in the soft moist soil and measured approximately 14 inches in length by 8 inches at the widest. The tracks were distinctly three toed, a trait often attributed to the Fouke Monster.

Corresponding reports have been cataloged in neighboring Texas, with a particularly high concentration in the counties nearest Fouke, such as Red River, Bowie, Cass, Marion, and Harrison. An encounter similar to the ones reported in Fouke happened to my friend Brad McAndrews while he was visiting his grandparents house in Marion County. (Marion County is just southwest of Arkansas along the Sulphur River Watershed.) I had heard details of Brad's encounter through mutual friends, but upon hearing the story from him firsthand, I had to seriously face the reality these reports represented. It's one thing to document reports from strangers, no matter how credible they seem or how positive their references are, but it is another thing entirely to hear an account from a friend and person who would have absolutely no reason to lie to you about such a thing. Brad is now a medical research biologist, but at the time he encountered a strange ape-like creature in the woods near his grandparent's home, he was simply an average ten-year-old kid who came upon something quite out of the ordinary.

It was around noon on a hot June day in 1989. Brad and his brother were staying at their grandparents' cabin which was located on a remote spot of land outside the town of Jefferson, Texas. On that day, Brad's grandmother and aunt offered to take the boys on a picnic, so the group of four ventured from the cabin down a path out past their old barn. Knowing of a spot that might be good for their picnic, Brad ran ahead to scout out the area. By now he was beyond sight of his grandmother and the others. When he arrived, he found that it was far more overgrown than he had remembered, so he turned and headed back towards the group.

As he headed back up the path, he heard a rustle in the woods. Something appeared to be coming directly toward him, so he stopped, thinking it might be a deer. As Brad waited anxiously, hoping to get a glimpse of a large buck, the cracking of twigs, pine saplings, and leaves became louder until finally an animal emerged from the woods. But this was no fine buck; it was something far more unexpected... and frightening. What emerged from the trees was an ape-like creature approximately seven-to-eight-feet tall with dark, reddish-brown hair and weathered looking skin. At first it was running on all fours, but upon seeing Brad, it stopped and rose up on two legs. It was then that Brad could see that it was, as he described it, "more human, than ape." In his own words:

> At about 15-20 yards to my right, it came into view and turned to its right. It was moving, on all fours, at what I would call a hustled pace, almost as if it was running from something. When it was at about 9-12 yards from me and about two feet off of the roadway, it used a rather large wooden fence post to hoist itself into a bipedal progression, using its left arm and hand. Its change in gait or posture did not result in a change in its speed. It then took a few running steps before passing behind a five foot sapling and stopping right before me. I could hear and feel the weight of this animal as it impacted the ground. When it reached me, it immediately stopped and squared its shoulders at me. I had never been so scared in my life, even to this very day. At only a distance of maybe 7-8 feet with nothing in between us, I completely froze and even held my breath hoping that maybe it wouldn't see/hurt me and just leave. I never felt that it was charging me, but rather that it happened upon me by accident. This creature

seemed to be just as startled to see me as I was to see it. His facial expression and body mannerisms told me that he was caught off guard and seemed very apprehensive as if it wasn't sure what to do. We locked eyes with each other for a number of seconds before he turned his upper torso back and to his right (as if looking over its shoulder), shifting his weight onto his right leg, as if considering to take off, only to turn back to me and lock eyes again, and repeating this twice more. He seemed to be curious about me, but looked like he wanted to get the hell out of there. He stood unobstructed right before my eyes for what seemed like an eternity, but probably only lasted maybe 8-10 seconds.

As he stood transfixed on the creature, Brad began to hear his younger brother calling his name as he came up the grassy pathway. At that point, the animal glanced in his brother's direction and suddenly sprinted off on two legs, using its hands to tunnel its way through the thicket of trees. Relieved, Brad turned and ran toward his brother as fast as his legs could carry him.

Sketch by Pete Travers based on Brad McAndrew's description.
(Courtesy of Pete Travers)

As an adult now, and having studied biology in college and in his ongoing profession, Brad was able to document his experience very thoroughly, reflecting on the nature of the animal:

First and foremost, I cannot explain enough how human-like this creature was. From its facial features and expression, body language, and walking/running gait, this creature was scarily human other than its other body characteristics. Even at my age, I could tell that this was an intelligent creature by the way it seemed to be assessing his circumstances. I also cannot express enough how fluid this creature was in shifting from a four legged "run" to a bipedal running gait. This creature does not have any problems, or handicap whatsoever, in its gait, ability to walk or run on two legs, or to progress from one to the other while moving at a high rate of speed. It was very much more human than "ape." I don't really like to use the word ape when describing it.

With understanding of my own height at the time and after taking that into consideration, I would estimate the creature to stand between 7 1/2 - 8 1/2 feet tall. He had very broad shoulders and heavy muscle structure. This was a massive creature that I would guess weighed in the vicinity of four to five hundred pounds, but I can honestly see it pushing 600 plus pounds. It did not appear shiny, but rather dull and coarse. I would guess the length to be about 3"-4". The hair covered most of the body, not including the face (with the exception of cheeks and jawline), forehead, palms of the hands, and the hair appeared much thinner on the chest and abdominal regions. Its head was cone-shaped slightly to the back of the head with the hair growing up and back—not growing down into the face. There was a forehead, as the hairline on the head did not extend down to its brow line (which was pronounced). Its neck was very thick and maybe short. I also believe that it may have had a restricted ability to swivel its head from side to side, as a human might do. Its eyes were a dark brown and its skin looked somewhat tough and weathered. I have difficulty describing the color of its skin. I can only describe it as an overly weathered and medium toned skin with a face that seemed more human than ape.

For obvious reasons, my friend's encounter, with its sobering detail and eerie description, sticks out in my mind. I often visited my extended family in the countryside of Polk County, not far from this incident, so I can easily envision the setting and circumstance. If I could be scared out of my wits at that age by simply watching *The Legend of Boggy Creek*, then I can only imagine how downright frightened I would have been to witness something like this in real life. Seeing a creature on film is one thing, but seeing it up close and personal must surely be something else entirely.

Not be left out, counties such Bossier, Caddo, and Claiborne in Louisiana have also seen the same trend of reports, reinforcing the idea that the natural network of waterways spreading out along the Sulphur River Watershed could be home to a population of these undiscovered animals or perhaps the offspring of a once-captive animal gone feral. Whether it is the same type of creature sighted near Fouke is impossible to determine - they all seem to be large and hairy - but regardless it does seem to present a heavy case for the presence of one or more strange beasts in the area. Strange beasts who might occasionally wander up the meandering path of the Sulphur River to the infamous Boggy Creek, and from there to the sleepy little town of Fouke.

DIMINISHING DOMAIN

In an attempt to dismiss the Fouke Monster sightings as isolated cases of mass hallucination or hysteria, some people point to the fact that the number of sightings has severely declined since the 1970s. Some people even claim that the sightings have stopped altogether. While I have shown that sightings definitely continued far beyond the monster's initial heyday, it is true that they have diminished in number over the years. There is no way of truly knowing why, but regardless, it doesn't provide enough ammo to debunk the whole affair. In fact, it might actually strengthen the case.

At the time of the initial rash of sightings during the 1960s and early 1970s, the landscape of Miller County was quite different than it is now. In those days, the land surrounding Fouke had not yet undergone the major deforestation that it has seen of late. Other than the small area carved out by Fouke's main street and Highway 71, most of the land in and around the town at the time was still heavily wooded. That's not to say that the area looks drastically different today, but there is definitely less woodland in the immediate vicinity than there once was.

The main two factors that have contributed to this deforestation have been the steady increase of open farmlands and the construction of Highway 549 that started early in the decade of the 2000s. The highway, which opened for traffic in December 2004, runs from Texarkana to Doddridge and takes a large bite out of lands formerly thick with trees. As in any case of major construction and land clearing, animals are forced to retreat further into undeveloped areas. This would, presumably, include any undocumented ape-like creatures that may dwell there.

If we look closely at the Fouke Monster sighting timeline, this diminishing wildlife domain seems to have had an effect on the reports of the monster. In the 1960s, most of the sightings took place near the outlying community of Jonesville, located near Mercer Bayou. Although Jonesville is still nestled in the dense Sulphur River woodland, it was even more isolated back in those days. Any creature moving through the landscape of the bayou could easily have happened upon one of the unassuming homes that dot the area and find itself face-to-face with humankind, as in the case of Mary Beth Searcy or Louise Harvin. Trappers and hunters were plentiful in the area back then and would be the most likely to stumble upon any such shadowy creature, as indeed they did. Sightings by the Crabtrees, Phyllis Brown, and Kenneth Dyas easily fit into this scenario, having hunted in and around the Mercer Bayou for most of their lives.

The encounters in the 1970s were closer to the town of Fouke, but this is not something that raises a major red flag. At the time, the spot where Boggy Creek crosses Highway 71, Willie Smith's soybean field, the Ford's rent house, and even the location on Oates Street closer to Texarkana were still sitting on the edge of vast wooded areas, and it was not uncommon to see wild animals at these locations. But over time, development in the community, and more dramatically, the conversion of woods to farmland has slowly lifted the veil of virgin timber and replaced it with the tarnish of civilization. As one would expect, the likelihood of seeing any large, reclusive animals near town was severely reduced.

As we move into the 1980s there are still some very impressive sightings, such as the one by Terry Sutton, but these are all limited to the Jonesville area or farther south in Mercer Bayou. Likewise in the early 1990s, two of the best sightings on record—where an unexplainable creature was seen by multiple witnesses—are confined to areas strictly on the outskirts of Fouke along Highway 71 or near the McKinney Bayou.

As farmland and other development continued around the immediate Fouke area, Jonesville stayed relatively remote. It is, therefore, not surprising that sightings in the later part of the 1990s were still being reported by Jonesville residents. The fact that no encounters closer to Fouke are on record, only strengthens the theory that the creature had permanently retreated to more remote areas to the south or southwest.

Three well-documented sightings by hunters are on record for the year 2000. In each case, the encounters took place near the Sulphur River south of Jonesville, as if to underscore the creature's slow progression southward to avoid the expansion of humanity. Following these sightings, survey and construction began for the Highway 549 project. During this time a few incidents were reported in the same area, but they seem to drop off dramatically as major construction got underway in 2002-2003.

As we approach the current era, even more clearing and development has taken place along the routes of Highway 549, State Highway 247 (Blackman Ferry Road) and other places, further reducing the chances that any such creature in his right mind would venture too far out of the dark reaches of Mercer Bayou or the bottomlands of the Sulphur River. I spoke to lifelong Jonesville resident Bob Sleeper during my research and he affirmed the slow evolution of landscape in the area. "As I boy, I used to walk all over this land from the Sulphur River to Texarkana," he told me. "I could disappear in there for weeks and never come across any other people. But now it's been cleared out so much, it's not the same."

Sleeper admitted that he has never seen the creature himself but agrees that sightings in the area would naturally be reduced as the land is cleared and the Texarkana metropolitan area expands. What was once an endless horizon of timber and swampland is slowly succumbing to the jaws of civilization. The land still has the power to hold secrets, but those secrets must be buried deeper into the recesses of shadow if they are to survive. Boggy Creek still holds Fouke in the grasp of its watery tendrils, but its namesake creature may well have abandoned his once favorite route.

TROUBLE WITH THREE TOES

It has been notoriously difficult to obtain hard evidence of the Sasquatch. As anyone who has watched one of the many television documentaries on the subject over the years well knows, all we have to show for the 50+ years of study on this phenomenon is a handful of grainy videos, an assortment of dubious photos, a few batches of inconclusive hair samples, and a warehouse full of footprint casts that vary widely in shape and size. These items, along with the eyewitness accounts, are the clues we have as to what the creature might be. Without a body there can never be any positive DNA samples, hair comparisons, or any other firm scientific proof that higher primates may exist on the North American continent. This lack of hard evidence certainly does not rule out the possibility of its existence, but it does little to conclusively solve the mystery, and oftentimes it raises more questions than it answers.

People often ask: "why doesn't anyone ever find a body?" or "how come nobody can get a clear picture?" as if the lack of these items somehow disproves the case. But you might just as well ask "when's the last time you ran across a dead bear in the road?" Probably never, even though bears are a very prominent species in North America. Likewise, when an unexpected animal accidentally presents itself for a few scant seconds, who has time to focus the camera for a glamour shot before it moves back into the trees? It's natural to ask these questions, but in the end they don't lead us any closer to an answer.

It seems that for all the questions and scholarly conjecture, there is also the matter of just how a person wishes to view the data. A different perspective can be applied to almost any scenario when solid scientific proof is not there to provide an iron-clad answer. If someone believes in the possibility of Sasquatch-type creatures, then they might see details in a photo that support that view. If they don't wish to embrace this concept, then they can just as easily see the photo in a totally different light. This is the way the human brain works.

The same goes for the countless footprints that have been found and attributed to an undocumented species of animal that, by all indications, has evolved to have a foot similar to that of man. The casts made from these footprints offer some of the best circumstantial evidence in support of Sasquatch, especially if we accept that some contain actual dermal ridge details (i.e., fingerprint lines), but still they cannot offer the ultimate answer to the question until there is a known sample with which to make a comparison.

In the case of the Fouke Monster, we do not have any videos or photos, blurry or otherwise. Nor do we have any definitive hair samples. The only somewhat solid evidence that the monster has possibly left behind has been the imprint of its foot. But like the Sasquatch phenomenon as a whole, these do not solve the mystery, and in fact, only throw more questions in the ring due to the unique properties of the Fouke Monster tracks.

The trouble with most of the prominent Fouke Monster tracks is that they show the creature as having three toes. While there have been some "Bigfoot" tracks found elsewhere across the United States that possess this three-toed configuration and even a four-toed configuration, in general, the Sasquatch is theorized to have five toes, which would be typical of any other higher primate, whether extinct or living in the world today. If the Fouke Monster is presumed to be one of these animals, and it does indeed have three toes, this may suggest that it is some species separate from any proposed population of Sasquatch creatures that may exist in the countryside. Understandably, this is something of a problem when trying to simplify the case for the Fouke Monster, since it implies that the creature might not even be a logical evolution of a primate—albeit a yet-to-be-discovered one—and makes it that much harder to believe there just might be something to all these sightings. John Green does a good job of expressing this problem in his book, *Sasquatch: The Apes Among Us*:

> Most show five toes, but about 20 percent of reports describe either four toes or three toes. Probably the proportion with less than five toes is not actually that great. The number of toes often is not mentioned in a footprint report, and it seems likely that when prints show three or four toes that would usually be remarked on, while five toes would be taken for granted.
> If there were just five-toed tracks and three-toed tracks and each type was of consistent shape, I would accept that as a clear indication of two different species. Since there are four-toed tracks as well, and the three-toes kind are very inconsistent in shape, I don't think such a conclusion would help much.

The foremost set of tracks attributed to the Fouke Monster were those found in Willie Smith's bean field on June 13, 1971. The footprints were measured to be 13.5 inches long by 4.5 inches wide and had a maximum distance of 57 inches between them. There was also the impression of some kind of digit several inches back from the big toe, but this was faint and at no time did it make a solid impression in the soft soil. The tracks definitely gave the impression of a large bipedal animal, which was puzzling to start with, but what made them even more mysterious is that the animal only had three main toes.

Some people who saw the tracks in person had doubts as to their authenticity. "I noticed that it stepped over the plants," reporter Jim Powell recalled. To him, it seemed a bit too convenient that the animal would have avoided stepping on any of the soybean plants. Even scientists weighed in on the debate. In an article appearing five days after the first report, in the June 17 edition of the *Texarkana Gazette*, archaeologist Frank Schambach from Southern State College in Magnolia, Arkansas, voiced his suspicions. "There is a 99 percent chance the tracks are a hoax," he stated for the record. Since the monster had been described thus far as some type of ape, he was quick to point out to the Fouke Monster fanbase, that: "All primates have five toes. This in my opinion would rule out any type of monkey or ape."

Some were not so quick to jump to conclusions either way, however. Rick Roberts noted the long stride measurements and the sheer number of tracks. "It would have been really hard to fake those," he told me, remembering his first impression when he saw the tracks in person as teenager. The game warden, Carl Gaylon, who examined the tracks at the time, could not make a definite ruling either, only that he had never seen tracks like that before. Constable Walraven, who was dubious about the monster at first, seemed swayed by the new evidence.

Smokey Crabtree weighed on the matter by telling the *Texarkana Gazette* that "the tracks found in the area in 1963 looked like the same print." This evidence also matched observations made at the Ford house six weeks earlier. According to the news report, "All that remained Sunday morning at the Ford house was several strange tracks—that appeared to be left by something with three toes…"

Another set of tracks thought to belong to the Fouke Monster was found by Orville Scoggins on November 25, 1973, after he witnessed a strange hairy animal walking on two legs across his field. These measured 5.5 inches in diameter and were 40 inches apart. Admittedly, it doesn't appear that these tracks were made by the same culprit who made the 13.5 inch tracks, but Constable Walraven was quoted as saying: "These are the same tracks that were found the first time the monster was sighted," indicating a connection.

Another vintage track cast is owned by Tom Zorn, who grew up in Fouke and has spent much of his life researching the Fouke Monster himself. The track casting was made in 1973 by Doris L. Brown, although I do not know the circumstances behind its discovery. It measures roughly 15 inches in length and 6.5 inches wide at the ball. The cast is extremely rough and does not show a very clear indication of the toe count one way or another.

More recently, people have reported finding tracks of the five-toed sort. These are obviously more in line with Sasquatch standards, but this only seems to confuse matters even more since they were not found in conjunction with any creature sighting, nor do they conform to the prominent belief that the Fouke creature has *three* toes.

The best of these five-toed samples—and the only specimen I have personally examined—was cast by Doyle Holmes in November 2004. The track, which represents the imprint of a left foot, measures 15 inches long by 8.5 inches wide at the toes and 5.5 inches wide at the ball. The photo here (and rather poor casting quality) does not do it justice, but upon examining it in person, it clearly has five distinct toe impressions. It also appears to have some kind of growth or bunion protruding from under the largest toe, which was consistent with all tracks found in the area.

Whatever or whoever created the tracks was not spotted at the time, but the circumstances surrounding their discovery suggest that they were not likely to be a hoax nor the impressions of any known animal in the area. I interviewed Mr. Holmes at length and also spoke to his son, Nathan, who initially spotted the tracks, and found them to be straightforward and sincere about their story. Holmes (mentioned previously in conjunction with the Giles "ape attack" of 1974), is a lifetime resident of Doddridge. He grew up hearing stories of the Fouke Monster and still lives in the area south of Mercer Bayou where he and his son often hunt and fish.

The discovery of the tracks took place a day or two after Thanksgiving in 2004. Holmes and his son were making the best of the holiday and decided to head up to what the locals call "Carter Lake" to hunt wild hogs. The swampy area can be extremely uncomfortable during the hot months, so a brisk hunt in the waning days of November is regarded as a welcome getaway.

As was their ritual, the father and son picked out some guns, lashed their canoe to the bed of their pickup, and drove the short distance up County Road 35 toward Carter Lake. When they reached the point where the road dead-ends into the swampy waters of Mercer Bayou, they parked the truck and traded their wheels for paddles.

Doyle and Nathan made their way down the water-gorged flatland, which meanders through a colony of standing cypress and an endless blanket of moss. About halfway to Carter Lake, the duo parked the canoe at the edge of the water between two old cypress trees and got out to look for signs of hogs. A short time later they came up on a large group of hogs and began to follow them along the bank near the water. As Nathan was looking for traces of hog

Strange track discovered in 2004 by Doyle and Nathan Holmes.
(Photo by the author)

tracks, trying to determine which way they had gone, he noticed a strange set of footprints leading out of the water toward a berry patch nearby. The tracks registered clearly in the moist soil, numbering about ten total, before they disappeared into vegetation. Hog tracks were one thing, but these looked very peculiar, so Nathan called his father over for a second opinion. Doyle inspected the tracks and was shocked at the size. They appeared human, with five toes, but they were much larger and had an oddly shaped foot.

The tracks were fairly fresh, so they followed them into the berry patch until they could no longer be seen. No other traces of their maker could be found—perhaps fortunately!—but still they were somewhat spooked by the discovery.

Without a way to cast the track, Doyle decided to mark the spot and return later. They headed out, wondering just what kind of animal had made the tracks. Inevitably, the Fouke Monster came to mind. Two days later, Holmes returned with cement, intent on preserving the tracks. Unfortunately many of them had begun to erode, and the best track had been partially stepped on a by a hog, but still he was able to get one good sample.

A rather spooky incident that occurred about seven months prior is relevant here. It took place around April of 2004, as Holmes was fishing in the same area. About one hour after sunrise he heard some movement and splashing on the nearby bank. Figuring it might have been a large hog, he peered into the trees trying to get a glimpse of it. Shortly afterward something came into view, but it definitely wasn't a four-legged hog. The thing, whatever it was, walked on two legs. Holmes could not get a clear visual of its upper body, due to the overhanging canopies of cypress, but he was able to see the figure from about the thigh down to the calf. There was no doubt it was a bipedal creature, but in the shadows, Holmes could not completely rule out that it was not a man in a pair of fishing waders. However, Holmes had been out on the water since before daylight and had not seen any signs of other people, not even a boat parked on the edge of the water. This was notable since the area was not accessible by any means other than by boat. He was miles from the road where the water began and could not imagine how another human could have been walking on the bank so far out, unless of course they lived in the swamp. Holmes had been coming here for years and was not aware of any resident in the area.

On one occasion, I had Doyle take me out to the area where he had seen the mysterious walker and where the tracks had been found. It was evident that the area would have been hardly accessible by a person on foot. To boot, the waters are filled with huge alligators that glided along near our canoe as if keeping an eye on us.

So in the end, what are we to make of all these troublesome tracks? Does the creature have three toes, five toes, or could these footprints be nothing more than misidentifications or hoaxes?

The trackway found in Willie Smith's bean field during the peak of Fouke Monster activity is the most dramatic example of "hard" evidence, but also perhaps, the most dubious. In a 1986 interview with the *Texarkana Gazette*, Fouke Mayor Virgil Roberts expressed his personal doubts about the tracks. "I feel like somebody walked out there with stilts on. I think they were trying to get attention. Maybe they thought they could make a little money on it," he told reporters. He went on to add: "I suspect there were several involved. I won't call their names because some people have passed on."

Although there is no proof of a hoax—even Mayor Roberts admits that no one has ever come forward to claim responsibility—it just might be that the tracks were man-made. Willie Smith himself seemed to be just the type of guy who could turn the Fouke Monster into a cash cow, if he only had a bit more sensational proof—or perhaps a track casting to sell at his service station! Smith and his family had several unverified run-ins with the monster, including an instance where he shot at the creature fifteen times but missed, so it seems very convenient that the mysterious creature just happened to leave his best set of prints right in Smith's freshly plowed field. It is also ironic that the creature did not damage any of the precious crop, although to be fair, Smith was leasing the land to Mr. Kennedy, so technically the crop was not his own. However, it is puzzling that if Smith did create them, why did he not make them more ape-like? He himself theorized that the monster was merely an escaped primate. Did he not know that apes have five toes? Or did he not care and simply base his design on the "evidence" found out the Ford house, which mentioned the curious three-toed tracks?

These questions are impossible to answer without further input from Smith, who is long deceased, or by further examination of the castings. Unfortunately, the original castings were destroyed when the Boggy Creek Café caught fire in the late 1970s. Copies are extremely hard to come by, if they even exist at all, and as a result I was unable to track one down.

If all the three-toed tracks are thrown out of court as hoaxes or mistakes, it might be easier to imagine the Fouke Monster as a natural primate, be it a Sasquatch or merely a bipedal orangutan or gorilla who has escaped the confines of a private owner. I spoke to wildlife biologist and university professor, Alton Higgins, who agreed: "In my opinion, no three-toed track has anything to do with the Sasquatch or any undocumented primate."

Alternately, we could theorize that because of a limited population, the three-toed foot represents some type of congenital deformation caused by inbreeding. An animal like this would surely have a limited population and, as such, a little cousin kissing might be necessary. However, this type of manifestation usually results in a more deformed looking foot all around, which is not the case here. The tracks found in the bean field show the foot to be fairly uniform except for the fact that it has three toes.

To be sure, I spoke to Dr. Jeff Meldrum, an Associate Professor of Anatomy and Anthropology at Idaho State University and the leading authority on Sasquatch track analysis. His conclusion was simple: "I have never seen a credible three-toed track."

MODERN TRAIL OF MYTHIC CREATURES

Cryptid research was not a very sophisticated activity back during the Fouke Monster frenzy. The locals did a commendable job of pursuing their mystery beast in the 1970s, but certainly the technology available to assist them was limited. For example, at the time the three-toed tracks were discovered, investigators called in one of the local women who had some crafting experience to pour Plaster of Paris into the depressions to make a mold. Investigators nowadays, whether law enforcement or amateur cryptozoologists, would come equipped with buckets of Hydrocal or other mixes of quality cement in order to cast the tracks. These substances are much more adept at picking up fine details such as skin folds, scars, or dermal ridge impressions. The soil (substrate) would also be analyzed for its density or other properties. Digital photos would be taken of the area and motion-sensitive cameras would be installed to capture an image of the perpetrator, should he return to the scene of the crime. With all the activity said to be going on, the Fouke area would have been a modern-day cryptozoologist's dream.

For those not familiar with the subject of *cryptozoology*, it is defined as the "study of hidden animals." Alleged hidden animals such as the Fouke Monster, Bigfoot, Chupacabra, lake monsters, and sea serpents fall into this category and are referred to as *cryptids*. Cryptozoology also includes creatures of the less monstrous sort, including the legendary black panther (whose existence is highly debated), giant salamanders, massive feral hogs, out-of-place kangaroos, and many more. It also encompasses a range of species once thought to be extinct or extirpated (i.e., killed off), which have been rediscovered or are believed to be still living. This category includes the Tasmanian tiger, dodo bird, giant moa, and the famous coelacanth, a 360-million year old fish hauled in by a fisherman off the coast of South Africa in 1938, much to the astonishment of scientists.

Other animals that would have once been categorized as cryptids include the giant panda and mountain gorilla. These are now familiar animals, but at one time they were thought to be figments of some overactive imaginations. Ironically, the mountain gorilla was initially described as a "man-like ape," much like the Fouke Monster, by the German explorer Captain von Beringe. These amazing primates were unknown until 1902 when he finally shot one near modern-day Rwanda and brought it back to Europe.

Today, with all the documentaries, books, and television shows, cryptozoology is much more widely known than it was back in the early Fouke Monster days. The locals didn't view the Fouke Monster as "cryptozoology," but as "something strange that their neighbor saw." It was more matter-of-fact than it was sensational.

Modern-day cryptozoologists employ other advanced equipment in their search for cryptids, including special low-light video cameras, night-vision goggles, thermal imaging devices, digital recorders, and portable "call-blasting" units. Call-blasting is the technique of broadcasting sounds such as primate howls, injured animal calls, alleged Sasquatch vocals, or even the sound of laughing children in an effort to illicit a response from an unknown animal. This is not unlike the technique Smokey Crabtree was using to entice the monster with a wounded rabbit call back in the 1960s. Modern call-blasting utilizes highly amplified audio to broadcast the sound over great distances. Whether this technique will help lure an undocumented animal from the shadows is uncertain, but there is no doubt that the forests echo with some very strange noises at night, some not so easy to identify.

Texas Bigfoot Research Conservancy investigator Jerry Hestand, who has a deep appreciation for the Fouke Monster phenomenon, told me once of an interesting experience he and some other researchers had while call-blasting south of Fouke. They were in the area of Thornton Wells, near where the coonhunter had his frightening encounter in the winter of 2000. The researchers were split up into several teams, one of them stationed at a higher point, which rose from the swampy floor of Thornton Wells. As the chilly night settled in, they began to blast sounds of "Bigfoot howls" every half hour or so into the surrounding darkness. After several rounds, something suddenly echoed the very same sound from deep in the bayou. It was so similar to the pre-recorded howl that some of the team thought another call-blast was being played without their knowledge. But this was not the case. They played the call again and the unseen animal responded, this time closer to the stakeout area. In fact, the animal seemed to be getting closer and closer.

As they continued to play the calls, one of the team members, who was parked near a small bridge, heard something large moving through the cypress trees around him, breaking limbs and splashing in the water. Much to his horror, a rock came hurling from the black reaches of the swamp and hit the side of his truck. By now, the man was nearly petrified. Hestand tried to spotlight the culprit, but nothing could be seen. The ruckus continued for awhile longer before the night finally went silent.

Oddly enough, while Hestand and I were conducting some interviews for this book in Fouke, a local hunter by the name of John Attaway mentioned that he had once heard animal howls, which reminded him of a howler monkey, in the same area. He did not have any prior knowledge of Hestand's experience at Thornton Wells. Not only that, but he and a friend were driving one evening right over the same little bridge in Thornton Wells where the ruckus had occurred. The men were moving very slowly, with the windows down, when they noticed a pungent odor. "It was so bad, I thought something was dead," Attaway told us. "Like somebody had killed a deer or walked off the road and dumped some hog guts or something." After smelling the stench, Attaway continued down the road for a short distance until he reached a boat ramp where he turned around. By the time they returned to the bridge, a mere two minutes later, the smell was completely gone. It was as if the source of the reeking odor had simply gotten up and walked off, which is pretty difficult to do if you happen to be a dead animal! Curious, the men plugged in a powerful spotlight and looked into the surrounding trees. They didn't see anything, so Attaway grabbed his gun and jumped out of the truck. He searched the area on foot, but could find no animals, dead or alive. Fearless as I know Attaway to be, it was probably fortunate that the animal had moved on. Otherwise, it wouldn't have lived one more smelly day, Fouke Monster or otherwise.

Was it mere coincidence that Attaway experienced something unexplainable in the very same spot that Jerry Hestand and his associates had?

In 2009, Hestand and other members of the Texas Bigfoot Research Conservancy made a significant contribution to the

"Swamp Stalker" episode of *MonsterQuest*, which focused primarily on the legend of the Fouke Monster. Cryptozoologist Loren Coleman provided historical background on the infamous creature, and the episode reenacted the more recent encounter by Stacy Hudson, which I mentioned previously. It also documented the efforts of two teams, comprised largely of TBRC investigators, working to solve the enduring mystery of the "Swamp Stalker" in the Arkansas-Louisiana-Oklahoma-Texas region.

One of the teams explored the spooky waters of the Sulphur River at night looking for any signs of mysterious creatures. They applied a high-tech reacting agent to the vegetation along the bank, which if touched, would reveal latent handprints when illuminated with UV light. Unfortunately nothing turned up during the filming, but it did illustrate some of the latest methods of field research. It is interesting to note that the scenes showing the team members kayaking through the water were shot near the Thornton Wells boat ramp, the very same area that had previously yielded promising results.

On one of my research trips to Fouke, I ran into a large group of investigators who were camping near Smith Lake west of the Mercer Bayou. The group was made up of various members of other dedicated cryptozoology research groups, including Mid-America Bigfoot Research Center, Southern Sasquatch Investigators, and the TEXLA Cryptozoological Research Group. They had come to Fouke for a weekend of research, bringing with them a staggering amount of equipment, everything from canoes and All Terrain Vehicles (ATVs), to high tech cameras and sound recording equipment. It was astounding to see just how many people were willing to put up with the swarms of mosquitoes, ticks, and swamp water to search for the alleged beast. At the end of the weekend, I asked if they had gotten any results. One of the guys played me a recording of a very interesting return call which came from the dark woods after a call-blast. It sounded like the roar of a large animal, perhaps a lion. I told them that a lion had indeed been seen nearby several years back, so you never know. As far as proof of the Fouke Monster, there was nothing conclusive, but this did not seem to phase the group. They promised to return in the future.

So it seems that all the modern technology and advanced gadgetry doesn't necessarily ensure success. But at the end of the day, it doesn't really matter if modern gadgetry proves the creature is or was ever real. What happened in and around the town of Fouke happened. Whether legend or flesh, it is still the Beast of Boggy Creek.

9. The Legend Lives On

Legacy At Large

If there's one thing I learned while writing this book, it's that the legend of the Fouke Monster is still very much alive around campfires, on television, in books, and on the internet. And since sightings are still being reported—some even more incredible than those that launched the beast of boggy creek to fame back in the 1970s—there is no doubt that the phenomenon will survive for as long as we have a desire for mystery or until the monster and its ilk are brought forth from the depths of their swampy home.

The Fouke Monster's media appearances continue to be on the rise with the boom of paranormal-related programs on television today. In addition to the History Channel's *MonsterQuest* coverage, the Fouke Monster was profiled on the top-notch independent documentary, *Southern Fried Bigfoot*, on the Travel Channel's *Weird Travels*, and incorporated into the semi-fictional premise of a *Lost Tapes* episode titled "Southern Sasquatch." Most recently, the legend of the Fouke Monster managed to slip into prime time with its coverage on the ABC show, *Wife Swap*.

Another notable testament to the creature's enduring legacy can be found in a public service ad created by the Keep Arkansas Beautiful Commission in 2007. In this 30-second television ad,[16] the character of the Fouke Monster is used to address the growing problem of litter. The tongue-in-cheek premise is effective and clever, in that it envisions the Fouke Monster as a legendary movie star 35 years after his most famous movie (i.e., *The Legend of Boggy Creek*) had been filmed. In keeping with this theme, the monster's hairline is receding, its teeth are capped, and it is generally shown to be older than it was in 1972 at the "height of his career." In the clip, the "monster" speaks out against the evils of littering, explaining that all kinds of animals live in the forest and none of them should have to put up with such clutter. The ad was run quite extensively, appearing on CNN, Discovery, Sci-Fi, and ESPN, as well as on local stations in Arkansas.

The aging Fouke Monster poses with creative director Chip Culpepper. (Courtesy of Mangan Holcomb Partners and Keep Arkansas Beautiful Commission)

The Fouke Monster has also inspired musical acts. There was the 45 RPM single "Fouke Monster" by Billy Cole and the Fouke Monsters from 1971. More recently, the extreme metal band, Troglodyte, paid tribute to the legend by titling their 2011 CD, "Welcome to Boggy Creek." The band's founder, Jeffery Sisson, is heavily influenced by Bigfoot lore and cites *The Legend of Boggy Creek* as a major influence. "Welcome to Boggy Creek" not only pays nod to Pierce's movie in the title, but the intro track was created as a sort of homage to the movie's spooky intro complete with swampy sounds, narration, and the inimitable scream of the monster.

A sympathetic Fouke Monster makes an appearance as the central character in two children's fiction books. The first, *The Return of the Fouke Monster,* was written and published by Ethel Wright in 1975. It can still be found in the local Texarkana college library. The second book, *The Night We Saw the Fouke Monster,* was written and published by Angelia Purvis in 2007. This one is still available online.

A far more frightening creature turns up in the 2011 indie film, *Boggy Creek*. Early buzz suggested that it was to be a remake of the original *Legend of Boggy Creek*, but as it turns out, it's a completely unrelated story set in the fictional town of Boggy Creek, Texas. Even so, the film, with its small-town setting and use of spooky swamp-scapes, obviously draws influence from Pierce's masterpiece, and one-by-one the cast of the movie is picked off by a ghastly hoard of Southern Sasquatch, which are infinitely more angry and violent than those in Pierce's film, as one would expect in today's horror market. The new *Boggy Creek* also shares common ground with *Creature From Black Lake*, as both films were shot on location at Caddo Lake.

The legend has also been successfully merchandised. A quick search of the internet will bring up various t-shirts based on *The Legend of Boggy Creek* movie or the Fouke Monster itself. The Monster Mart in Fouke continues to sell its own t-shirts and hats, and Smokey Crabtree is happy to sell fans a souvenir or two at his bookstore or at any of his Bigfoot-related conference appearances.

Various monster merchandising.
(From the personal collections of Lyle Blackburn and Loren Coleman)
The stuffed doll was handmade by the late Karen Crabtree.
The dolls sold at the Monster Mart from about 2002-2007.

In town, hardly a day goes by that some visitor doesn't stop into one of the local establishments to ask about the monster. Some of the locals may just smile and laugh, but there are still plenty of folks who are willing to tell a quick story or two, or tell about the time their grandparents helped make a movie. The subject of the Fouke Monster seems to have touched almost every resident, whether directly or indirectly through friends and relatives. To some outsiders it may seem like a "fringe" subject, but to most of the locals, this is a part of everyday life in Fouke, something that runs through their blood like the web of creeks that runs through the surrounding woods.

It would be hard to live in Fouke for too long without becoming aware of the town's notorious claim to fame. I spoke to one newcomer, a teenager, who had recently moved to Fouke with his family. I asked him if he had heard of the Fouke Monster prior

During the 1980s, the town incorporated the monster into the theme of an art fair.
(Courtesy of the Miller County Historical Society)

to moving there. He had not, but he said that within a week of living there, his mother went out to rent a copy of *The Legend of Boggy Creek* so they could get up to speed on the local history. They enjoyed the campiness of the movie, but he admitted that now, when he rides through the local cornfields, he often throws a wary glance over his shoulder just in case something might be lurking in the shadows of the treeline. "It's always in the back of my mind," he told me.

It might indeed be wise to keep an eye on the woods around Fouke, even today. A sighting of a mysterious creature by two locals in May 2010 suggests that something still lurks in their midst. The two locals, Michael and Liz Rowton, were both "non-believers" when it came to the Fouke Monster. Liz had been employed at the Monster Mart convenience store for more than seven years, so she had heard more stories from locals and tourists than most anyone in Fouke, but even this did not convince her that it might be a real flesh and blood animal. However, after a startling encounter in the spring of 2010, her mind was forever changed.

Liz had been working the late shift at the store, so by the time her husband Michael picked her up it was approaching 11:00 p.m. It had been raining that day, but by the late evening it had dissipated into a fine mist that covered Fouke in dreary haze. The couple left the Monster Mart and headed north on Highway 71 until they reached the County Road 22 turnoff, not too far from the property where the Fords were allegedly attacked by a creature so long ago. Michael turned right onto Co. Rd. 22 and proceeded down the long winding pavement that cuts its way through the mass of thick trees on either side. The fine mist glistened as the headlights blazed across the wet macadam, but visibility was not a problem. After traveling a mile or so, they passed a small cluster of oil wells in a clearing on the left side of the road. As they approached, Michael noticed something ahead, something squatting in the grassy area between the edge of the road and the wall of trees. Caught in the headlights, the thing stood up and quickly ran across the road a mere 20 yards in front of them. It was so sudden that Liz was looking down at the time and did not see the creature. However, she knew something unusual had happened when Michael abruptly slowed down and craned his head in the direction of the strange runner.

Alarmed, Liz asked her husband to explain… immediately. In a tone of utter disbelief, he simply said: "I think I saw the Fouke Monster." The look on his face told her that he was not playing a joke. She urged him to turn around so they could get a better look. Michael slowed down and did a 180 in the road. The headlights spilled into the trees where he had seen the creature run. Scanning the lighted area for any sign of the creature, they caught a glimpse of its shadowy outline as it stood just inside the cover of tall pines. Michael eased the car forward for a better look, but the animal, or whatever it was, quickly disappeared from sight.

By now, Liz thought better of any further investigation, feeling a surge of fear replace the initial whirlwind of excitement. The brief glimpse of the shadowy figure had given her the creeps. Knowing that it could still be watching them, she insisted they get out of there and only come back when it was daylight to investigate. Michael agreed as he spun the car around and hit the gas. He had no idea what he had just seen, but it was definitely something he could not explain.

The couple described the creature as being approximately seven feet tall, covered in dark hair, and running upright on two legs. Michael had the best view, but in the scant few seconds he saw it running, he was not able to make out any clear facial features. They did not "believe" in the Fouke Monster, but they had no choice to consider the possibility at that point. The shape, the size, and the way it moved ruled out most rational possibilities. Unless it was some joker in a suit, willing to sit in the rainy darkness and risk life and limb to run out in front of passing cars, there seemed to be no other explanation.

I interviewed Liz at length one day, as she went over the details of their strange sighting. I could clearly see the anxiousness on her face as she recounted the events of that evening. It was obvious that whatever they saw had a lasting impact on the young girl. She told me that they returned the following day to search for footprints or other signs of the creature but came up empty-handed. The

Black birds sit atop one of the many oil well structures in the Fouke area.
(Photo by Chris Buntenbah)

pavement is lined by gravel and grass, so even though the soil was moist, there were no signs of its passing.

In the days immediately following the sighting, the couple was reluctant to tell anyone. After years of hearing stories at the Monster Mart, but always dismissing them with a smile, Liz now found herself in the opposite corner. Who would believe her and Michael? But in the end, they saw what they saw, and no amount of teasing would make them believe any differently. Eventually they began to tell their closest friends, and that's how their encounter ended up being public knowledge. The legend was indeed alive in Fouke and running on two legs through the shadowy woods.

Conclusion

On my final trip to Fouke, I stumbled upon a perfect example of how the legend thrives beyond the confines of its namesake town. It was an unseasonably warm day in March when I pulled to a stop in the gravel parking lot of the quaint Miller County Museum with a couple of friends. We got out, stretched a bit, and then headed inside. The museum was rather busy with several locals chatting inside and a family of tourists gathered around the humble Fouke Monster display case to marvel at the various photos and relics.

I greeted the curator, Frank McFerrin, and after a minute or two, Frank introduced me to the visiting family, telling them of my book project. I shook hands with a nicely dressed man who introduced himself as Jerry Garcia, certainly an unforgettable name to anyone acquainted with another famous hairy guy from the Grateful Dead. Not to be outdone, I announced myself as Lyle Blackburn, the monster's *biographer*. After a laugh and a few questions, Mr. Garcia introduced me to his wife and daughter, stating that they had come to Fouke because of their interest in Bigfoot and in particular, the Fouke Monster. The family was on vacation and since they would be passing near Arkansas, they agreed that it would be worthwhile to detour through Fouke. So after driving approximately 450 miles from the San Antonio area of Texas, they spent the day trying to find out more about the monster as they explored the countryside where it is said to roam.

I later saw the family taking photos of the mural at the Monster Mart, and it made me realize just how powerful the allure of the legend really was, drawing people from miles around to the little town of Fouke. After meeting the Garcia family, I knew that my work would be appreciated by many others.

Later that afternoon, with my friend and fellow researcher Jerry Hestand, I walked around the very spot where the Ford incident occurred almost 40 years ago to the day. The house was long gone,

The property where the Ford incident occurred.
The house is no longer there.
(Courtesy of the Miller County Historical Society)

but this was still ground zero, the launch pad for a legend that has worked its way like a thorn into the greater lore of Arkansas. I felt a bit like the narrator in *The Legend of Boggy Creek* as he returned to the old house where he and his mother had first heard the monster's horrible wail from the depths of the soggy bayou. It had been a long time since I had first seen the movie come to life on that big drive-in screen in the 1970s, but I can still remember how its haunting scream affected me that night.

In some ways I wasn't surprised that my life had brought me to the very spot where the old Ford house had once stood, trying to imagine what had gone down that night so long ago. Essentially I was just another visiting tourist from Texas, an outsider to Fouke, but somehow I felt that I belonged, that it was my job to tell the whole story of this shadowy beast, which had grabbed my imagination as a kid and had never really let go. I looked at Jerry and we laughed at our mutual childlike fascination as we stood there. We were still in the clutches of the mysterious beast.

Nobody will ever be able to prove whether or not the Fords actually came face-to-face with an unknown creature that night in

1971, but it doesn't matter. Mysteries are not about the end result, they are about thrills, chills, and *possibilities*. It is the pursuit of possibility and the exhilaration of the chase that rewards us, not just the end result. We may never find the answer to the question of whether man-apes, UFOs, and sea serpents truly exist, but that will not stop us from constantly re-examining the facts and trying to conquer the unknown because that is what we do as humans. We have an innate fascination for things we don't understand, and that is a good thing. Mysteries are the catalysts of our progress, whether they lead to something earth-shattering or simple personal satisfaction.

The Fouke Monster may seem fantastic, but it is difficult to dismiss all the reports. Eyewitness accounts are notoriously unreliable and some of the details are inconsistent, but regardless, there are many reports that just can't be written off so easily. What about the multiple instances where an upright hairy creature was seen crossing the road by half a dozen people? What about the policemen who reported a similar sighting 20 years earlier? How about the Grammy Award-winning musician who saw a man-like animal running down a desolate part of Highway 71 years before the public knew of the legend? Or the sighting by Terry Sutton, a well-respected young man in the community of Jonesville who saw it walk by his family's pond one evening? Having spoken to these people, and finding many of them absolutely sincere and credible, I am left with the opinion that they certainly saw *something*. And that seems to be the general consensus of those knowledgeable with the Fouke Monster phenomenon.

So as the sun sets over the lonely stretches of Boggy Creek, what remains is a story that thrills us, entertains us, and perhaps reminds us that we should never give up the search for new answers, even when we think we know everything about our world. We may not ever find proof of an unknown species of man-like animals living in this remote corner of Arkansas, but as long as someone somewhere continues to speak the name of the Fouke Monster, the legend will surely live on.

ENDNOTES

1   I later became aware of the so-called goatman of Lake Worth, Texas, which is much closer to my childhood home. However, as I child I was not aware of this monster since its brief wave of sightings were reported prior to the 1970s.

2   It is possible that humans migrated to the area earlier, but due to erosion and decomposition archeologists have yet to find any proof that Arkansas was inhabited prior to this.

3   The newspaper identifies Bobbie Smith as Willie Smith's daughter-in-law, but this was incorrect. Local sources informed me that Bobbie was most likely a niece-in-law.

4   Mr. Phillips served in Miller County law enforcement from 1969 until he retired in 2006. He is intimately familiar with the Fouke Monster case.

5   Since writing this book, the Monster Mart has changed ownership. The latest owner, Denny Roberts, has made an effort to pay better tribute to the town's infamous legend. If you visit the Monster Mart in the future, you may find a much improved Fouke Monster display than I found upon my initial research trips there.

6   I discovered J.F. Shaw while doing research on this book. Coincidently, my great-grandfather Malcolm Shaw lived and died in Panola County, Texas, which is a mere 93 miles from Fouke. Intrigued, I spent some time researching my family tree to determine if I might be related to the man. I could not find a recent link, but I did trace my family back to immigrants from Scotland, which is significant since Shaw's father was of Scottish-Irish decent. Perhaps in the future I can locate a common ancestor, but for now it shall remain a curious footnote.

7   The Below bridge is located about seven miles south of Fouke on County Road 8. It goes over the McKinney Bayou. It is technically east of Jonesville, but can still be considered as part of the same general area.

8   Many news sources cite the year as 1963, but according to *Smokey and the Fouke Monster* the year was 1965.

9   Smokey had always dreamed of providing his family with their own home in the countryside. After purchasing the land, Smokey built his home by hand, then created a beautiful lake on the property by clearing the land

and engineering a levee himself. The lake is officially recognized on the Arkansas map as Crabtree Lake.

10　The movie credits attribute the role of Mary Beth's sister to Mary B. Johnson.

11　Not to be confused with the set decorator named Charles Pierce. The Internet Movie Database (www.imdb.com) lists his set decoration credits under Charles B. Pierce's profile, but this is incorrect. This is one reason Charles B. Pierce always stressed the "B." initial in his name to avoid confusion.

12　Allegedly, this was a three-toed track.

13　As with the sightings that are particular to the Crabtree family, I recommend reading Smokey's book for more details about what transpired after the skeleton came into Smokey's possession.

14　Wayne Scoggins is not related to the previously mentioned Raymond, Orville, or Jerry Wayne Scoggins of Fouke. This is just a rather confusing coincidence.

15　Perhaps not so coincidentally, there have been reports of monkey howls coming from the Ouachita Mountain region of Arkansas and Oklahoma. These mountains are considered to be a hotspot for Bigfoot sightings.

16　The commercial can be viewed at www.foukemonster.net

APPENDIX:
STRANGER THAN FICTION

REEL TO REAL

For those who like to separate fact from fiction, I have included a breakdown of scenes in Charles Pierce's *The Legend of Boggy Creek*, mapping them to the equivalent "true" reports by individuals. Being that it's a movie, people tend to think that many of the scenes were just made up, but this is not the case. Of course, some scenes are embellished for dramatic purposes—it is entertainment, after all—but nearly everything was based on real life reports.

This list is broken down by the *minute:second* where the scene starts in the movie:

2:34   SCENE: Young boy is running across a field on his way to Fouke to report that his mother had seen a "big hairy monster."
= Based on an incident which occurred in the 1960s in which a mother sent her seven-year-old boy on a two-and-a-half-mile run to the town of Fouke, where he informed the landlord that they had seen a large hairy creature approach their home.

9:35   SCENE: Willie Smith shoots at the monster from his porch.
= Based on Mr. Smith's own report in which he claims to have shot at the monster in 1955.

10:02 SCENE: John P. Hixon, owner of the Apache Ranch, describes how he and his son saw the monster running on two legs across their field. The monster appeared to be

218

wounded.

= Unknown source. No records of an Apache Ranch can be found.

10:30   SCENE: John W. Oates describes how his two prize hogs were killed and dragged into the woods. He suspects the monster.

= Based on Mr. Oates' own account in which two of his hogs were killed and subsequently carried off into the woods in 1971. According to Oates, he laid both dead hogs — one weighing 80 pounds and the other 70 pounds — at the edge of his land one evening. The next morning he discovered that both had been picked up and carried off (not dragged). He could still see the imprint of the hog's hair in the soft dirt of a gopher mound where they had lain. The movie theorized that it could have been the work of the monster, but the "monster" was never suspected in real life. Oates lived approximately 5 miles from Boggy Creek.

13:35   SCENE: Fred Crabtree encounters the monster while hunting in the woods.

= Based on Fred's own report in which he claimed to have seen the monster sometime in the 1960s.

15:00   SCENE: James Crabtree encounters the monster near a creek while hunting in the woods.

= Based on James' own report in which he claimed to have seen the monster along a creek bank in 1955. He believes it may have been washing itself in the creek just before he startled it.

16:55   SCENE: Mary Beth Searcy, her sister, and her mother see the creature from their window one night in the 1960s. Their cat is found dead the following morning.

= Based on the Searcy's own report in which they claim the monster stalked around the house one evening after dark. The death of the cat, however, was added to the film for dramatic effect.

23:37   SCENE: Teenage deer-hunter hears dogs barking and runs out to investigate, taking his shotgun. He encounters the monster, fires twice, and runs for the house, dropping his gun in the process.

= Based on a claim made by fourteen-year-old Kenneth Dyas that he had seen the monster in 1965 while deer hunting. Frightened, he shot at the creature and ran for the house. One month prior to Dyas' sighting, fourteen-year-old Lynn Crabtree also had a similar encounter. Lynn fired three times before running. A subsequent investigation did not locate any blood, so it is uncertain whether the monster was wounded or not.

26:16   SCENE: A search party—including men on foot and horseback, as well as bloodhounds—is formed to look for the monster.

= Following Lynn Crabtree's horrifying encounter in 1965, men started showing up at Smokey Crabtree's house offering assistance to hunt down the creature. Nearly 20 locals, armed with guns, horses and tracking dogs, scouted the area but were unable to find any solid evidence. In 1971, concerned by the new rash of incidents, Miller County Sheriff Leslie Greer organized a new search party to seek out the monster in and around Boggy Creek. Nothing conclusive was ever found.

28:29   SCENE: One member of the search party, a man on horseback, encounters the monster while in the woods alone. Startled, his horse throws him from the saddle.

= Based on the report made by Jimmy Kornett, who participated in the original hunt on the Crabtree property (in response to Lynn Crabtree's sighting). While riding his horse through the bottoms, Kornett claimed that a "creature" passed nearby, causing his horse to throw him. He could not say whether the animal he saw was a bear or something else.

38:00   SCENE: Herb Jones, a man who has lived by himself in the Sulphur River Bottoms for more than 20 years, expresses his disbelief in the monster.

= Herb Jones was a real man who lived in a place known as Little Mound on the Mercer Bayou.

41:05 SCENE: A couple driving along Highway 71 at night see the monster run across the road.

= Based on the report made by Mr. and Mrs. D.C. Woods Jr., and Mrs. R.H. Sedgass in May of 1971, in which they saw the monster run across the highway while returning from Shreveport one night.

41:18 SCENE: Monster scares chickens and cows.

= Many Fouke area locals suspected the monster was responsible for killing some of their livestock. However, no actual reports of chicken coop raids can be found.

42:08 SCENE: O.H. Kennedy discovers a series of strange, three-toed footprints one morning in Willie Smith's soybean field located near Boggy Creek. Local officials and wildlife experts are called in to investigate.

= This incident actually occurred in June of 1971. Mr. Yother Kennedy discovered a series of mysterious footprints in a soybean field, which he leased from Willie Smith. The tracks originated in the woods at one corner of the field and traveled about 150 yards before disappearing into the trees on the other side. Experts could not determine what kind of animal (if it was animal) had made the tracks.

47:00 SCENE: Bessie Smith and her children encounter the monster at the edge of their property near the woods.

= Based on an alleged sighting by Bessie Smith. No written records of this account can be found in the newspapers, so it is uncertain how accurately this scene is portrayed. But either way, local sources confirm that Ms. Smith did report a sighting.

49:16 SCENE: Charlie Walraven sees the monster run across the road while driving at night.

= Based on a claim made by Charlie Walraven in which he saw a creature run across County Road 9 south of Fouke.

50:09 SCENE: Nancy and some friends see the creature from

their window.

= Based on an incident that occurred in 1971 in which three girls—Chris Rowton, Sherrie Johnson, and Christine Worrell—were alone one night in a trailer home. The trailer was located on the east side of Highway 71 near Boggy Creek and Willie Smith's bean field. They heard noises all night as something stalked around on the porch. They never saw anything, but the following day they found "large greasy tracks" left by some unknown animal.

54:50   SCENE: The monster stalks around Howard Walraven's trailer home and is suspected of killing his dog.

= Based on a claims and conjecture made by Howard Walraven.

55:57   SCENE: The Ford incident.

= Based on the incidents which occurred in late April/early May of 1971 at a rent house where the Fords were living at the time. The subsequent newspaper reports by Jim Powell of the *Texarkana Gazette* inspired Charles B. Pierce to make the movie and essentially ignited the Fouke Monster craze of the 1970s.

CHRONICLE OF SIGHTINGS NEAR FOUKE

This list includes reported sightings of hairy bipedal man-ape creatures within 50 miles of Fouke, Arkansas. This is not a comprehensive list of all sightings that have occurred in Arkansas or the surrounding states, but only those which could possibly be attributed to an unknown cryptid that may (or may *have*) inhabited areas near Boggy Creek. (Sources are included unless the story was reported to or verified by the author directly.)

1908   Near Fouke, Arkansas – Woman is reported to have seen a creature fitting the description of the "Fouke Monster" when she was 10 years old.

1916   Summer / Near Wright-Patman Lake in Texas (Knight's

Bluff) / 19 miles from Fouke – While driving at night (from town to home) in a mule-drawn wagon, family sees a tall, hairy figure emerge from the woods and stand in full moonlight. It was described as "taller than a man and covered with long dark hair." The creature walked slowly toward them on two legs, all the while shrieking and motioning angrily. When the father fired at it with his rifle, the creature ran back into the woods. About Paranormal (http://paranormal.about.com) and reported privately to TBRC (www.texasbigfoot.org).

1932   Near Fouke, Arkansas (Jonesville) – Man sees a hairy, man-like animal from his porch. It quickly moved out toward the fence on the property and disappeared into the woods.

1946   Near Fouke, Arkansas – Woman sees a strange animal in the field by her house. According to the local sheriff, she said "it looked kind of like a man, and walked like a man, but she didn't think it was a man." *Texarkana Gazette*.

1947   Winter / Mooringsport, Louisiana (Caddo Lake) / 40 miles from Fouke – Two boys see a hairy, manlike creature peer into the window of their house one night. The creature was very tall, since the window was nearly eight feet from the ground because the house sat on tall piers. TBRC.

1955   Near Fouke, Arkansas (Mercer Bayou) – Experienced hunter/trapper sees a large gorilla-like creature while fishing in a river. It appeared to be washing its hands before it walked off on two legs into the trees.

1955   Near Fouke, Arkansas (Jonesville) – Two adult residents see a hairy, man-like creature cross the road one evening while driving near their home. They noted that it "walked like a man, but was too hairy to be a man."

1955   Fouke, Arkansas – Man sees a large, hairy ape-like creature near his house on Boggy Creek. He shot at it 15 times with a rifle, but apparently missed. *Victoria Advocate*.

1960s   Summer / North of Hughes Springs, Texas / 46 miles from Fouke – Two sisters see an animal with "lots of hair that looked somewhat like an ape but was walking tall and upright" through their bedroom window one night. TBRC.

1964   Near Fouke, Arkansas (Jonesville) – Teenage girl sees a tall, hairy, two-legged creature lurking in the yard of her rural home one evening.

1965   Near Fouke, Arkansas (Mercer Bayou) – Hunter sees a seven-foot tall, hairy man-like animal in the woods. As he approached, the creature disappeared into the thick underbrush.

1965   Near Fouke, Arkansas (Jonesville) – Teenage boy encounters a hairy man or ape-like beast near a lake on his family's property. It was described as seven-foot tall with "reddish-brown hair about four inches long all over its body. It stood upright like a man, but had extra-long arms." As the thing approached, he fired at it with his shotgun three times and then fled. *Sasquatch: The Apes Among Us* (Hancock House Publishing, 1971).

1965   Near Fouke, Arkansas (Jonesville) – Another teenage boy sees the alleged monster in the woods while deer hunting. He fired his gun at the creature before running away.

1965   Near Fouke, Arkansas (Jonesville) – Woman deer hunting sees a dark, hairy ape-like figure running on two legs along a pipeline.

1965   August / Between Prospect and Lodi, Texas / 38 miles from Fouke – Boy sees a "large hairy man or beast" while walking on a rural road one afternoon. The creature reportedly chased the boy for a short distance before disappearing again into the woods. *Marshall News Messenger.*

1966   Near Fouke, Arkansas – School bus driver sees the alleged monster cross the road early one morning. *Smokey and the Fouke Monster* (Day's Creek Production Corporation, 1974).

1966   Various occurrences / Near Fouke, Arkansas (Jonesville) – Smokey Crabtree reports that the alleged monster, on several occasions, approached his house at night, was heard screaming, or would disturb the dogs (although it never harmed any animals). *Smokey and the Fouke Monster*.

1967   Near Fouke, Arkansas – Man sees hairy, man-like creature run across a road while driving on a rural road near his house.

1967   Spring / Near Fouke, Arkansas – While driving late one night, a teenager (who later became a Grammy-award winning musician) and his cousin see a hair-covered, bipedal creature running along Highway 71. They were residents of Texarkana and had never heard of the Fouke Monster or Bigfoot. The teenager didn't realize what he seen until years later when he saw *The Legend of Boggy Creek*.

1968   Near Fouke, Arkansas (Jonesville) – Woman sees an upright, ape-like creature twice in the early morning hours. It was first seen eating scraps from a hog pen and later seen licking food remnants from a dog's bowl. She described it as having hair all over its body with a flat, brown nose and canine teeth like a dog or baboon.

1969   Near Alleene, Arkansas / 49 miles from Fouke – Several witnesses report seeing a "half-man, half-ape looking creature" while on a camping trip. They also found evidence of three-toed tracks. BFRO (www.bfro.net).

1970s   *Exact year unknown* / Near Fouke, Arkansas – Farmer sees a large, man-like creature covered in scraggly hair walking in his pasture. As told to director Charles B. Pierce.

1970s   *Exact year unknown* / Near Fouke, Arkansas – Man and his son see the alleged monster running on two legs across their field. The creature appeared to be wounded. As told to director Charles B. Pierce.

1971   May / Fouke, Arkansas – Bobby Ford, Don Ford, Elizabeth Ford, and Charles Taylor claim to see a seven-foot tall, hairy creature several times during the night as it approached their rent house and stuck a hand through the window. Investigators later found evidence of scratches on the house and three-toed tracks in the yard. *Texarkana Gazette*.

1971   May / Fouke, Arkansas – Three well-respected local
        residents see a "monkey-like" creature with long arms
        and dark hair run across Highway 71 near Boggy Creek.
        Texarkana Daily News.

1971   June / Near Fouke, Arkansas – Three people see a "tall
        hairy creature with red eyes" during the evening hours. It
        appeared to be "squatting" on an embankment at the edge
        of the woods, across the street from their home. *Texarkana
        Gazette.*

1971   June / Near Fouke, Arkansas – Child reports seeing a
        "monster in the woods" near his house. Same location as
        the previous report. *Texarkana Gazette.*

1971   June / Fouke, Arkansas – A trail of three-toed tracks are
        discovered in Willie Smith's soybean field. The trackway
        appeared to have been made by a bipedal creature walking
        upright. *Texarkana Gazette.*

1971   June / Fouke, Arkansas – Woman sees a six-foot-tall, hairy
        creature walking through the woods near Willie Smith's
        soybean field during the late evening hours. *Texarkana
        Gazette.*

1971   June / Fouke, Arkansas – Two local residents witness a
        creature fitting the monster's general description "slouch"
        across a gravel road in front of their car early one morning.
        Texarkana Gazette.

1971   June / Fouke, Arkansas – Two men from Kansas report
        seeing a strange, two-legged animal standing on the side of
        the road. *Texarkana Gazette.*

1971   June / Fouke, Arkansas – Several women and children
        report seeing an "ape-like creature" in the area. The
        women had come to the area to view the tracks in the
        soybean field. *Texarkana Gazette.*

1971   November / Near Sarepta, Louisiana (Bayou Bodcaw) /
        35 miles from Fouke – Squirrel hunter sees six-foot-tall
        creature with "reddish-brown hair all over its body about
        the length of a deer's hair" emerge from some brush near a
        creek. The witness first thought it was a person but realized

it was a man-like animal after it crawled up the creek bank and looked back at him. After a few seconds it "jogged up the creek until it was out of sight." BFRO.

1973   November / Near Fouke, Arkansas – Local man, known to be a skeptic, sees a four-foot-tall, hairy bipedal creature creep across his field early one morning. *Texarkana Gazette*.

1974   Near Fouke, Arkansas (Carter Lake/ Mercer Bayou) / 7 miles from Fouke – Two brothers are driving their truck toward home one evening when an ape-like creature jumps into their trailer and begins thrashing about. The men stopped and the creature fled into the woods.

1978   October / Bowie County, Texas (Sulphur River) / 35 miles from Fouke – Teenage boy hears something making noise in the woods at night and finds large footprints a few days later. His friends also claim to have seen what looked like "a large man in an Eskimo coat." TBRC.

1981   February / Near Fouke, Arkansas – Teenage boy sees a tall, man-like creature with scraggly fur and long arms walking on two legs near the family's fishing pond.

1981   July / Near Fouke, Arkansas – Teenage boy sees a hairy, man-like creature splashing through the Sulphur River.

1984   June / Near Caddo Lake, Texas / 39 miles from Fouke – While riding in a car at night, woman sees a "big hairy creature" run across the road and jump a fence. TBRC.

1984   October / Bowie County, Texas (Sulphur River) / 35 miles from Fouke – While walking on a dirt road in misty fog, woman smells a "wet animal smell" and sees a figure standing at the edge of a creek. The figure was estimated to be seven feet tall with long arms, and was covered in dark hair or fur. TBRC.

1985   May / Near Mena, Arkansas / 40 miles from Fouke – At sunset, a couple driving north on Highway 41 near the Red River encountered a seven- or eight-foot-tall figure near the road. It had "reddish-brown long hair and it stood upright with very long arms." TBRC.

1986   November / Wright-Patman Lake, Bowie County, Texas /
        28 miles from Fouke – Late in the afternoon, a deer hunter
        sees a six- or seven-foot-tall creature with dark brown hair
        all over its body. It had a foul smell and took very long
        steps and "walked with a slight stoop." TBRC.

1987   Fall / Near Benton Lake, Texas / 42 miles from Fouke
        – In mid-afternoon, a hunter sees a "large, hair-covered
        animal." It walked on two legs, and he was able to view
        the animal for two minutes from less than 50 yards away.
        TBRC.

1987   November / Near Oil City, Louisiana (adjacent to Black
        Bayou) / 36 miles from Fouke – Hunter sees a six-foot-tall
        creature with shaggy, dark-brown hair walking upright on
        the edge of a levee. The creature approached the man
        until it got within 100 yards, at which point it noticed
        the eyewitness. It paused for three to four seconds before
        running off over the levee into the woods. BFRO.

1988   Spring / Near Karnack, Texas / 42 miles from Fouke –
        Before dawn, while in a deerstand, a hunter sees a barrel-
        chested creature walking upright within 75 yards. TBRC.

1988   Fall / Near Waskom, Texas / 50 miles from Fouke – At
        dawn, while in a deerstand, a hunter sees a tall creature
        with a hairy head, face, and neck striding over a hill on its
        hind legs. TBRC.

1989   June / Northwest of Jefferson, Texas / 43 miles from
        Fouke – Boy has face-to-face encounter with large bipedal
        creature covered in hair with human-like face. He saw
        the creature from a distance of 9-12 yards away for 8-10
        seconds.

1990   October / Near Atlanta, Texas / 24 miles from Fouke –
        While driving at night on Highway 59, a group sees a tall,
        hairy creature run across the road and into the woods.
        TBRC.

1990   October / Near Fouke, Arkansas – Two men from
        Oklahoma smell a pungent odor while driving on Highway
        71. They pull over and see a tall, man-like creature

covered in shaggy black hair, running toward the south bank of the Sulphur River. The creature was described as walking "upright just like a human, not like a bear or gorilla." *Texarkana Gazette*.

1991   November / Karnack, Texas / 52 miles from Fouke – A large skeleton is discovered in the woods by two hunters. The specimen is completely intact except for a missing head, tail, and claws. Suspecting that the bones might belong to the alleged Fouke Monster, the hunters take the bones from the property and give them to Smokey Crabtree.

1992   October / Near Fouke, Arkansas – Six people (five young men in a car and one man in a semi truck) traveling on FM 134 around 11:00 p.m. stop their vehicles to watch a large, hairy animal walk upright across the road. The creature was estimated to be seven feet tall with a bulky frame covered in thick, bushy hair. It was clearly silhouetted in the headlights as it crossed the pavement and walked into a field where it disappeared into the darkness. All six witnesses got out of the vehicles and discussed the amazing sighting before leaving the scene.

1993   Near Fouke, Arkansas – Five people (two men in a truck, a couple in a car, and a man in a semi truck) traveling on Highway 71 at approximately 2:00 a.m. stop their vehicles to observe a large, hairy animal standing on the side of the road (much like the sighting of October of 1992). The creature stood approximately 100 yards from the witnesses and was described as being ape-like, yet it moved on two legs like a man. After observing the growing crowd of people for a few moments, it turned and darted into the woods, running fast on two legs. The witnesses discussed the sighting before leaving the scene. *Too Close to the Mirror* (Day's Creek Production Corporation, 2001).

1993   Near Fouke, Arkansas – Four hours after the previous sighting and only a half mile away, a woman sees an animal fitting the description of the "Fouke Monster" cross

County Road 10 in the early morning hours. *Too Close to the Mirror.*

1993   November / Near Linden, Texas / 32 miles from Fouke – Boy sees a huge figure standing on the front deck of his home. It was described as being much taller than six feet and broad-shouldered. TBRC.

1994   November / Wilton, Arkansas / 39 miles from Fouke – An individual encounters a seven-to-eight-foot-tall, black figure along a railway near dusk. Acting territorial, "it growled and had an odor worse than a skunk." TBRC.

1995   April / Hooks, Texas / 27 miles from Fouke – After nightfall, a landowner and friend see an ape-like creature about seven-foot tall standing behind a barn. After they shine a light on it, the creature retreated into the woods. TBRC.

1996   December / Woodlawn, Texas / 50 miles from Fouke – While traveling at night, man and his son see a large, hairy creature kneeling near a creek a few feet from the road. TBRC.

1997   Summer / Near Wright-Pittman Lake, Cass County, Texas / 18 miles from Fouke – After midnight, while camping, four boys witness a "figure leaning out of the tree line" from about 100 yards away. They heard crashing sounds and a loud scream. TBRC.

1997   September / Near Fouke, Arkansas –While working on his car in the driveway, a man sees a "large, dark-brown creature" standing in the woods watching him. He approached the creature and observed it for several minutes. Deciding to retrieve his gun, he went back to his house. When he returned, the creature was gone, but he eventually spotted it again, sitting in woods. He continued to observe the creature until dusk made it too hard to see. GCBRO (www.gcbro.com).

1998   July / Near Fouke, Arkansas – A woman babysitting some children sees a "very large, hairy creature." She watches them from the edge of the woods while on a morning walk.

Frightened, she quickly returned to the house. GCBRO.

1998   February / Near Vivian, Louisiana (Iron Bridge) / 27 miles from Fouke – While walking down a secluded oil field road at dusk, a man sees very large, hairy animal run quickly across the road. The creature had black fur and a "gorilla-like head," and made a loud moaning sound as it ran. TBRC.

2000   January / Near Fouke, Arkansas – Hunter observes a "hairy man-like creature approximately eight-foot tall" walking in a wheat field during the late afternoon. The wheat field was approximately two miles from the Sulphur River along Boggy Creek. The creature was also described as having a musky smell. TBRC.

2000   October / Near Fouke, Arkansas – A bowhunter sees a "big black figure" approaching his tree stand before dawn. The creature reached up and grabbed his arrow with a "large black hand, very similar to the hand of a gorilla." TBRC.

2000   Winter / Near Fouke, Arkansas – A coonhunter encounters a hairy "creature of immense size" late at night while hunting with his dog. It made a hissing sound and splashed water on the hunter while making deep, throaty sounds. TBRC.

2001   Near Fouke, Arkansas – Hunter comes face-to-face with a large, ape-like creature while hunting near Mercer Lake.

2001   Near Fouke, Arkansas (Jonesville) – Two young girls encounter a "hairy monster" while riding bikes in the evening hours. When they returned home and told their mother, she contacted the authorities.

2003   March / Near Harleton, Texas / 50 miles from Fouke – A driver sees a large hairy creature standing within 100 yards of the road at dusk. According to the witness, the creature "crossed the highway in three strides." TBRC.

2004   August / Near Atlanta, Texas / 27 miles from Fouke – While driving at night on Highway 56 towards Atlanta, a motorist sees a "huge beast" with six-to-eight-inch-long brown hair stepping over a barbed-wire fence. TBRC.

2005   October / Near Caddo Lake, Texas / 33 miles from Fouke –
        Hunter encounters a large, hair-covered figure just before
        dawn. The creature stood more than six feet tall and was
        observed from a distance of only 10 yards. TBRC.

2006   May / Lewisville, Arkansas / 20 miles from Fouke – At
        sundown while fishing, a man sees an eight-foot-tall
        creature covered in orange-brown hair 50 yards away. It
        made a sound described as cross between "a cow bellow
        and a panther." TBRC.

2007   June / Hempstead County, Arkansas / 30 miles from Fouke
        – Two teenagers hear loud knocking in the trees while
        fishing in a pond near a large pasture. At dusk, a large,
        hairy, ape-like creature ran from the woods into the pond
        and "began swinging its arms violently." The creature was
        covered in long, reddish-brown hair with a thick neck and
        very long arms. GCBRO.

2010   May / Fouke, Arkansas – While driving late one evening,
        a married couple sees a large, hairy creature run on two
        legs across the road. Light precipitation was falling, but
        the figure was visible in the headlights as it made its way
        across the pavement and into the trees beyond. The couple
        turned the car around to get a better look but was only able
        to get one final glimpse of the creature standing in the
        trees before it disappeared in the darkness.

Bibliography

Books

Bord, Janet and Colin. *Bigfoot Casebook Updated: Sightings and Encounters from 1818 to 2004.* Pine Winds Press, An imprint of Idyll Arbor, 2006 (first published in 1982).

Chandler, Barbara Overton and J. Ed Howe. *History of Texarkana and Bowie and Miller Counties Texas-Arkansas.* J. Ed Howe Publisher, 1939.

Coleman, Loren. *Bigfoot: The True Story of Apes in America.* New York: Paraview Pocket Books, 2003.

Crabtree, J.E. Smokey. *Smokey and the Fouke Monster.* Fouke: Days Creek Production Corporation, 1974.

Crabtree, J.E. Smokey. *Too Close to the Mirror.* Fouke: Days Creek Production Corporation, 2001.

Green, John. *Sasquatch: The Apes Among Us.* Washington/Vancouver: Hancock House, 1978.

Halpin, Marjorie and Michael Ames. *Manlike Monsters On Trial: Early Records and Modern Evidence.* University of British Columbia, 1980.

Matthews, Rupert. *Bigfoot: True-Life Encounters with Legendary Ape-Men.* London: Arcturus Publishing, 2008.

Miller County Historical Society. *Fouke, Arkansas In Word and Picture Volume I 1891-1941.* Fouke: Miller County Historical Society, 1991.

Miller County Historical Society. *Fouke, Arkansas In Word and Picture Volume II 1942-1192.* Fouke: Miller County Historical Society, 1993.

Shinn, Josiah H. *Pioneers and Makers of Arkansas.* Little Rock: Genealogical Publishing, 1967.

MAGAZINES

Fuller, Curtis. "I See by the Papers." *Fate* March 1972: 24-28.
Jones, Mark and Teresa Ann Smith. "Has Bigfoot Moved to Texas?" *Fate* July 1979: 30-36.
Wooley, John. "The 'Boggy' Man." *Fangoria* #165 Aug. 1997: 13-18.
"Animals: Battle of the Species" *Time* 12 Nov. 1951.

NEWSPAPERS

Bischof, Greg. "'Boggy Creek' remains apparently Siberian tiger." *Texarkana Gazette* 11 Dec. 1991.
Brown, Donald C., Jr. "The Arkansas Monster Season - A Double Feature, No Less." *Dispatch* 28 Oct. 1971.
Davis, Robert. "Fouke lives with monsters and memories." *Texarkana Gazette* 27 Mar. 1988.
Ford, Patricia C. Letter, The Fords tell 'true' Fouke monster story. *Texarkana Gazette* 22 Sep. 1972.
Gelder, Austin. "Hunter's remote camera captures picture of mountain lion." *Arkansas Democrat* 23 Aug. 2003.
Hutson, Landel. "Little Rock radio station boosts reward for monster." *Texarkana Gazette* 25 June 1971.
Hutson, Lindel. "'Fouke Monster' — What's 'out there' no laughing matter." *Texarkana Gazette* 27 June 1971.
Lacy, Bettie. "Monster put Fouke on the map." *Texarkana Gazette* 25 May 1976.
Powell, Jim. "Fouke family terrorized by hairy 'monster.'" *Texarkana Gazette* 3 May 1971.
Powell, Jim. "'Monster' is spotted by Texarkana group." *Texarkana Daily News* 24 May 1971.
Powell, Jim. "'Fouke Monster' seen again." *Texarkana Gazette* 25 May 1971.
Powell, Jim. "'Fouke Monster' legend lives on." *Texarkana Gazette* 23 Jan. 1972.
Powell, Jim. "Fouke 'monster' is labeled possible sub-human creature." *Texarkana Gazette* 28 Sep. 1972.
Powell, Jim. "'Fouke Monster' movie will open at local theater tonight." *Texarkana Gazette* Aug. 1972

Powell, Jim. "Strange footprints spotted again." *Texarkana Gazette* 2 Feb. 1974.

Robyn/Staff Writer. "The Fouke Monster: Science Fiction or Science Fact?" *Texarkana Gazette* 25 Feb. 2001.

Ross, Margaret. "Fouke 'Monster' Had Look-Alikes." *Arkansas Gazette* 27 June 1971.

Smith, George. "Fouke put on map." *Texarkana Gazette* 22 July 1973.

Staff Writer. "Unknown Creature Sighted in Arkansas: Wild Man of the Woods." *Memphis Enquirer* 09 May 1851.

Staff Writer. "The Wild Man Again." *Hornellsville Tribune* 08 May 1856.

Staff Writer. "In search of monster." *Texarkana Gazette* May 1971.

Staff Writer. "'Creature' Attacked, Victim Says." *Arkansas Gazette* 4 May 1971.

Staff Writer. "Officers gird for Fouke 'monster hunt.'" *Texarkana Gazette* 7 May 1971.

Staff Writer. "'Monster' sighted; no tracks found." *Texarkana Gazette* 4 June 1971.

Staff Writer. "Meandering monster missed once again." *Texarkana Gazette* 6 June 1971.

Staff Writer. "Creature's prints found by officers." *Texarkana Gazette* 7 June 1971.

Staff Writer. "'Monster' tracks found." *Texarkana Gazette* 15 June 1971.

Staff Writer. "Archaeologist says 'tracks' are hoax." *Texarkana Gazette* 17 June 1971.

Staff Writer. "Bounty is offered for 'Fouke Monster.'" *Texarkana Gazette* 23 June 1971.

Staff Writer. "On the Rialto screen." *Southeast Missourian* 12 Apr. 1973.

Staff Writer. "Family files suit asking 'Boggy Creek' show halt. *Texarkana Gazette* 20 Apr. 1973.

Staff Writer. "Legendary Monster Becomes Money-Maker." *The Victoria Advocate* 23 Aug. 1973.

Staff Writer. "$10,000 is offered to capturer of Fouke 'Monster.'" *Texarkana Gazette* 16 Sep. 1973.

Staff Writer. "Fouke Monster Reward Offered." *The Victoria Advocate* 16 Sep. 1973.

Staff Writer. "Nonbeliever at Fouke Spots 'Monster' in Cow Pasture." *Arkansas Gazette* 27 Nov. 1973.

Staff Writer. "No search for the 'monster' is planned." *Texarkana Gazette* 27 Nov. 1973.

Staff Writer. "Fouke 'Monster' law suit is dismissed upon request." *Texarkana Gazette* 8 Dec. 1973.

Terry, Bill. "Tales of Arkansas Monsters." *Education Bureau Gazette* Oct. 1972

Thibodeau, Sunni. "The Fouke Monster 30 Years Later." *Texarkana Gazette* 24 June 2001.

Watkins, Warren. "Killing animals for personal protection is legal, commission says." *The Daily Citizen* 22 Dec. 2006.

West, Bob. "'Monster' hunters facing 'bear facts.'" *Texarkana Gazette* 22 June 1971.

Wexler, Laura. "Darkness on the Edge of Town." *The Washington Post* 23 Oct. 2005.

Wicker, Bill. "'Monster' sighting reported." *Texarkana Gazette* 26 Nov. 1973.

Williamson, Jim. "Arkansas delivers monster of a message." *Texarkana Gazette* 19 Sep. 2007.

Yount, Sheila. "Boggy Creek 'monster' still stalks Fouke folks." *Arkansas Democrat* 22 May 1989.

ONLINE ARTICLES

Colyer, Daryl and Alton Higgins. "Bigfoot/Sasquatch Sightings: Correlations to Annual Rainfall Totals, Waterways, Human Population Densities and Black Bear Habitat Zones." Texas Bigfoot Research Conservancy website (http://www.texasbigfoot.com/index.php/about-bigfoot/articles/67-ecological-patterns).

DeBlack, Thomas A. "Civil War through Reconstruction, 1861 through 1874." The Encyclopedia of Arkansas History website (http://encyclopediaofarkansas.net/encyclopedia/entry-detail.aspx?entryID=388)

Dougan, Michael B. "Transportation." The Encyclopedia of Arkansas History website (http://encyclopediaofarkansas.net/encyclopedia/entry-detail.aspx?entryID=399).

Hendricks, Nancy. "Texarkana (Miller County)." The Encyclopedia of Arkansas History website (http://encyclopediaofarkansas.net/encyclopedia/entry-detail.aspx?entryID=934).

Lancaster, Guy. "Red River." The Encyclopedia of Arkansas History website (http://encyclopediaofarkansas.net/encyclopedia/entry-detail.aspx?entryID=2650).

Loewen, James. "Sundown Towns." The Encyclopedia of Arkansas History website (http://encyclopediaofarkansas.net/encyclopedia/entry-detail.aspx?entryID=3658)

Rowe, Beverly J. "Miller County." The Encyclopedia of Arkansas History website (http://encyclopediaofarkansas.net/encyclopedia/entry-detail.aspx?entryID=790).

Simpson, Ethel C. "Otto Ernest Rayburn (1891-1960)." The Encyclopedia of Arkansas History website (http://encyclopediaofarkansas.net/encyclopedia/entry-detail.aspx?entryID=3003)

Staff writer. "Civil War Battle Summaries by State." National Park Service website (http://www.nps.gov/hps/abpp/battles/bystate.htm)

U.S. Fish and Wildlife Service (Northeast Region). "U.S. Fish and Wildlife Service concludes eastern cougar extinct" U.S. Fish and Wildlife Service website (http://www.fws.gov/northeast)

WEBSITES

Arkansas Department of Parks and Tourism <http://www.arkansas.com>

Bigfoot Field Researchers Organization <http://www.bfro.net>

Caddo Lake History Page <http://www.caddolake.com/history.htm>

Circus Historical Society <http://www.circushistory.org>

National Park Service <http://www.nps.gov>

Encyclopedia of Arkansas History and Culture <http://encyclopediaofarkansas.net>

Gulf Coast Bigfoot Research Organization <http://www.gcbro.com>

Internet Movie Database, The <http://www.imdb.com>

Phantoms & Monsters <http://naturalplane.blogspot.com>

Texas Bigfoot Research Conservancy <http://www.texasbigfoot.com>

ACKNOWLEDGMENTS

Special thanks to the following friends and family members who contributed so much to this book. Without them, it would not have been possible.

My beautiful wife, Sandy.

Jerry Hestand, Frank McFerrin, Denny Roberts, and Rick Roberts.

Thanks also to…

Conor Ameigh, Dave Alexander, Rusty Anderson, John Attaway, Syble Attaway, Cathy Bennett, Chris Buntenbah, Jimmy Clem, Dave Coleman, Loren Coleman, Daryl Colyer, Smokey Crabtree, Eerie Eric, Carl Finch, Duane Graves, Frank Garrett, Dave Hall, Bobby Hamilton, Mackey Harvin, Alton Higgins, Doyle Holmes, Peanut Jones & the boys, Chris Kuchta, Stacia Langenheder, Justin Meeks, Dr. Jeff Meldrum, Keenan McClelland, Larry Moses, H. L. Phillips, Nick Redfern, Robert Robinson, Chris Rowton, Liz Rowton, Bob Sleeper, Amanda Squitiero, Kathy Strain, Ken Stewart, Lloyd Sutton, Terry Sutton, Robert Swain, Danny Vail, Sean Whitley, Craig Woolheater, Bigfoot Field Researchers Organization, Gulf Coast Bigfoot Research Organization, Mid-America Bigfoot Research Center, Texas Bigfoot Research Conservancy, the Monster Mart, and the *Texarkana Gazette*.

And to Jim Powell, who covered the initial Fouke Monster sightings in the *Texarkana Gazette*. Without his work, the creature may have lived forever in obscurity.

ABOUT THE AUTHOR

Growing up in Texas, Lyle Blackburn became fascinated with the legends, lore, and sighting reports of alleged real-life monsters. He is a frequent contributor and cryptozoology advisor to *Rue Morgue* magazine, one of the leading horror media publications in print today. He is also the founder and frontman for the rock band Ghoultown. Over the last decade, Ghoultown has released six albums, toured extensively in both the U.S. and Europe, and has appeared on several horror movie soundtracks. Lyle currently lives near Dallas, Texas, where he enjoys a day off now and then.

For more information, visit the following websites:

www.monstrobizarro.com

www.foukemonster.net

www.ghoultown.com

CPSIA information can be obtained at www.ICGtesting.com
Printed in the USA
BVOW09s1042291214

381072BV00011B/256/P